SAFE WORK
in the 21ST CENTURY

Education and Training Needs for
the Next Decade's Occupational
Safety and Health Personnel

Committee to Assess Training Needs for
Occupational Safety and Health Personnel in the United States

Board on Health Sciences Policy

INSTITUTE OF MEDICINE

NATIONAL ACADEMY PRESS
Washington, DC

NATIONAL ACADEMY PRESS • 2101 Constitution Avenue, NW • Washington, DC 20418

NOTICE: The project that is the subject of this report was approved by the Governing Board of the National Research Council, whose members are drawn from the councils of the National Academy of Sciences, the National Academy of Engineering, and the Institute of Medicine. The members of the committee responsible for this report were chosen for their special competences and with regard for appropriate balance.

Support for this project was provided by the National Institute for Occupational Safety and Health, the Occupational Safety and Health Administration, the U.S. Department of Veterans Affairs, the National Institute of Environmental Health Sciences, the American Association of Occupational Health Nurses, and the American Academy of Industrial Hygiene. This support does not constitute endorsement of the views expressed in the report.

Library of Congress Cataloging-in-Publication Data

Safe work in the 21st century : education and training needs for the next decade's occupational safety and health personnel / Committee to Assess Training Needs for Occupational Safety and Health Personnel in the United States, Health Sciences Policy Division, Institute of Medicine.
    p. cm
Includes bibliographical references.
ISBN 0-309-07026-0
    1. Industrial hygiene. 2. Industrial safety. I. Institute of Medicine (U.S.). Committee to Access Training Needs for Occupational Safety and Health Personnel in the United States.
RC967.S215 2000
616.9'803—dc21   00-055005

Cover photograph: Window Washer on the Empire State Building. Photographer: Earl Dotter.

The serpent has been a symbol of long life, healing, and knowledge among almost all cultures and religions since the beginning of recorded history. The serpent adopted as a logotype by the Institute of Medicine is a relief carving from ancient Greece, now held by the Staatliche Museen in Berlin.

"Knowing is not enough; we must apply.
Willing is not enough; we must do."
—Goethe

INSTITUTE OF MEDICINE

Shaping the Future for Health

THE NATIONAL ACADEMIES

National Academy of Sciences
National Academy of Engineering
Institute of Medicine
National Research Council

The **National Academy of Sciences** is a private, nonprofit, self-perpetuating society of distinguished scholars engaged in scientific and engineering research, dedicated to the furtherance of science and technology and to their use for the general welfare. Upon the authority of the charter granted to it by the Congress in 1863, the Academy has a mandate that requires it to advise the federal government on scientific and technical matters. Dr. Bruce M. Alberts is president of the National Academy of Sciences.

The **National Academy of Engineering** was established in 1964, under the charter of the National Academy of Sciences, as a parallel organization of outstanding engineers. It is autonomous in its administration and in the selection of its members, sharing with the National Academy of Sciences the responsibility for advising the federal government. The National Academy of Engineering also sponsors engineering programs aimed at meeting national needs, encourages education and research, and recognizes the superior achievements of engineers. Dr. William A. Wulf is president of the National Academy of Engineering.

The **Institute of Medicine** was established in 1970 by the National Academy of Sciences to secure the services of eminent members of appropriate professions in the examination of policy matters pertaining to the health of the public. The Institute acts under the responsibility given to the National Academy of Sciences by its congressional charter to be an adviser to the federal government and, upon its own initiative, to identify issues of medical care, research, and education. Dr. Kenneth I. Shine is president of the Institute of Medicine.

The **National Research Council** was organized by the National Academy of Sciences in 1916 to associate the broad community of science and technology with the Academy's purposes of furthering knowledge and advising the federal government. Functioning in accordance with general policies determined by the Academy, the Council has become the principal operating agency of both the National Academy of Sciences and the National Academy of Engineering in providing services to the government, the public, and the scientific and engineering communities. The Council is administered jointly by both Academies and the Institute of Medicine. Dr. Bruce M. Alberts and Dr. William A. Wulf are chairman and vice chairman, respectively, of the National Research Council.

COMMITTEE TO ASSESS TRAINING NEEDS FOR OCCUPATIONAL SAFETY AND HEALTH PERSONNEL IN THE UNITED STATES

JAMES A. MERCHANT (*Chair*), Dean, College of Public Health, University of Iowa, Iowa City

LINDA HAWES CLEVER, Chairperson, Department of Occupational Health, California Pacific Medical Center, San Francisco

RUTH HANFT, Independent Health Policy Consultant, Charlottesville, Virginia

RONALD KUTSCHER, Retired Associate Commissioner, Office of Economic Growth and Employment Projections, Bureau of Labor Statistics, Washington, D.C.

JAMES A. OPPOLD, Occupational Safety and Health Consultant, Raleigh, North Carolina

M.E. BONNIE ROGERS, Director, Occupational Health Nursing Program, School of Public Health, University of North Carolina at Chapel Hill

SCOTT SCHNEIDER, Director, Occupational Health and Safety Program, Laborers' Health and Safety Fund of North America, Washington, D.C.

MARTIN SEPULVEDA, Vice President, Global Occupational Health Services, IBM Corporation, Somers, New York

ROBERT C. SPEAR, Professor of Environmental Health Sciences, School of Public Health, University of California at Berkeley

LOIS E. TETRICK, Professor of Psychology, Department of Psychology, University of Houston

NEAL A. VANSELOW, Chancellor Emeritus, Professor of Medicine Emeritus, School of Medicine, Tulane University

M. DONALD WHORTON, Occupational Medicine Practitioner, President, M. Donald Whorton, Inc., Alameda, California

Board on Health Sciences Policy Liaison

MARK R. CULLEN, Professor of Medicine and Public Health, Occupational and Environmental Medicine Program, Yale University School of Medicine, New Haven, Connecticut

Study Staff

FREDERICK J. MANNING, Project Director
ALDEN B. CHANG II, Project Assistant

Institute of Medicine Staff

ANDREW POPE, Director, Board on Health Sciences Policy
HALLIE WILFERT, Administrative Assistant
CARLOS GABRIEL, Financial Associate

INDEPENDENT REPORT REVIEWERS

This report has been reviewed in draft form by individuals chosen for their diverse perspectives and technical expertise, in accordance with procedures approved by the National Research Council's Report Review Committee. The purpose of this independent review is to provide candid and critical comments that will assist the institution in making its published report as sound as possible and to ensure that the report meets institutional standards for objectivity, evidence, and responsiveness to the study charge. The content of the review comments and draft manuscript remain confidential to protect the integrity of the deliberative process. We wish to thank the following individuals for their participation in the review of this report:

JOHN M. DEMENT, Associate Professor, Division of Occupational and Environmental Medicine, Department of Community and Family Medicine, Duke University Medical School

BERNARD D. GOLDSTEIN, Director, Environmental and Occupational Health Sciences Institute, University of Medicine and Dentistry of New Jersey-Robert Wood Johnson Medical School

JOSEPH LADOU, Professor of Occupational Medicine and Toxicology, Division of Occupational and Environmental Medicine, University of California at San Francisco

JANE A. LIPSCOMB, Associate Professor, School of Nursing, University of Maryland, Baltimore

EDWARD B. PERRIN, Professor Emeritus, Department of Health Services, Center for Cost and Outcomes Research, University of Washington

GORDON REEVE, Corporate Epidemiologist, Health Care Management, Ford Motor Company, Dearborn, Michigan

JONATHAN D. ROSEN, Director, Health and Safety Department, New York Public Employees Federation, Albany

DAVID ROSNER, Professor of History and Public Health, and Codirector, Program in History of Public Health and Medicine, Mailman School of Public Health, Columbia University

GAVRIEL SALVENDY, Professor of Industrial Engineering, Purdue University

DAVID TOLLERUD, Director, Center for Environmental and Occupational Health, C.P. Hahnemann University School of Public Health, Philadelphia

While the individuals listed above provided many constructive comments and suggestions, responsibility for the final content of the report rests solely with the authoring committee and the Institute of Medicine.

vii

Acronyms and Abbreviations

AAIHN	American Association of Industrial Nurses
AAOHN	American Association of Occupational Health Nurses
AAOM	American Academy of Occupational Medicine
ABET	Accreditation Board for Engineering and Technology
ABOHN	Accreditation Board for Occupational Health Nurses
ABIH	American Board of Industrial Hygiene
ABPM	American Board of Preventive Medicine
ACGIH	American Conference of Governmental Industrial Hygienists
ACGME	Accreditation Council for Graduate Medical Education
ACOEM	American College of Occupational and Environmental Medicine
ADA	Americans with Disabilities Act
AEP	Associate Ergonomics Professional
AHC	academic health center
AIHA	American Industrial Hygiene Association
AMT	advanced manufacturing techniques
AOMA	American Occupational Medicine Association
ASP	Associate Safety Professional
ASSE	American Society of Safety Engineers
BCPE	Board of Certification in Professional Ergonomics
BCSP	Board of Certified Safety Professionals

CAE	Certified Associate Ergonomist
CE	continuing education
CEA	Certified Ergonomics Associate
CEAP	Certified Employee Assistance Professional
CEPH	Council on Education in Public Health
CEU	continuing education units
CHCM	Certified Hazard Control Manager
CHFEP	Certified Human Factors Engineering Professional
CHFP	Certified Human Factors Professional
CIE*	Certified Industrial Engineer
CIH	Certified Industrial Hygienist
CMI	computer-managed instruction
COHN	Certified Occupational Health Nurse
COHN-S	Certified Occupational Health Nurse-Specialist
CPE	Certified Professional Ergonomist
CSP	Certified Safety Professional
DOE	U.S. Department of Energy
DOJ	U.S. Department of Justice
EAP	employee assistance program
EAPA	Employee Assistance Professional Association
EFS	Educational Field Services program
EPA	Environmental Protection Agency
ERC	Education and Research Center (Educational Resource Center until 1999)
GATT	General Agreement on Tariffs and Trade
GDP	gross domestic product
HMO	health maintenance organization
HWWT	hazardous waste worker training
IHIT	Industrial Hygienist in Training
IOM	Institute of Medicine
IVD	interactive video disc
JIT	just-in-time inventory control
MAC	maximum allowable concentrations
MCW	Medical College of Wisconsin
MSHA	Mine Safety and Health Administration
MWT	Minority Worker Training program

NAFTA   North American Free Trade Agreement
NIEHS   National Institute of Environmental Health Sciences
NIOSH   National Institute for Occupational Safety and Health
NLN   National League for Nursing

OEM   occupational and environmental medicine
OM   occupational medicine
OSH   occupational safety and health
OSHA   Occupational Safety and Health Administration
OSHAct   Occupational Safety and Health Act

RPE   Registered Professional Engineer

Acknowledgments

George W. Anstadt
Eastman Kodak Co.

Roger L. Brauer
Board of Certified Safety
Professionals

Thomas Bresnahan
American Society of Safety
Engineers

Ann Brockhaus
Organization Resources
Counselors, Inc.

Leo Carey
National Safety Council

Scott Clark
University of Cincinnati

Jerome Congleton
Texas A&M University

Ann Cox
American Association of
Occupational Health Nurses

Ann Cronin
National Institute for
Occupational Safety and Health

Cathy Cronin
Occupational Safety and Health
Administration Training Institute

Gregory DeLapp
Employee Assistance Professional
Association

Don Ethier
American Industrial Hygiene
Association

Julia Faucett
University of California at San
Francisco

Adam Finkel
Occupational Safety and Health
Administration

Bruce G. Flynn
Washington Business Group on
Health

Roy Gibbs
United States Department of
Energy

Manuel Gomez
American Industrial Hygiene
Association

Larry Grayson
National Institute for
Occupational Safety and Health

William Greaves
Medical College of Wisconsin

Colonel Mark Hamilton
Office of the Deputy
Undersecretary of Defense

Eugene Handley
American College of Occupational
and Environmental Medicine

Larry Hardy
American College of Occupational
and Environmental Medicine

Joseph Hughes, Jr.
National Institute for
Environmental Health Sciences

Sharon Kemerer
American Board for Occupational
Health Nurses

W. Monroe Keyserling
University of Michigan

Bernadine B. Kuchinski
National Institute for
Occupational Safety and Health

Tom Leamon
Liberty Mutual Insurance Co.

Tom MacLeod
Mine Safety and Health
Administration

Michael S. Morgan
University of Washington

Royce Moser, Jr.
University of Utah

Frances M. Murphy
U.S. Department of Veterans
Affairs

Julie B. Norman
University of Montana

John Olson
University of Wisconsin-Stout

Nico Pronk
Health Partners

Jonathan Rosen
New York State Public Employees
Federation

Linda Rosenstock
National Institute for
Occupational Safety and Health

Karl Sieber
National Institute for
Occupational Safety and Health

Rosemary Sokas
National Institute for
Occupational Safety and Health

Tim Stephens
University of North Carolina at
Chapel Hill

John T. Talty
National Institute for
Occupational Safety and Health

Victor Toy
American Association of
Industrial Hygienists

Lawrence W. Whitehead
University of Texas-Houston

Jerry Williams
American Society of Safety
Engineers

Samuel Wilson
National Institute of
Environmental Health Sciences

Contents

LIST OF TABLES, FIGURES, AND BOXES

Tables

Figures

Boxes

Executive Summary

ABSTRACT. Work in the United States has changed in the decades since the Occupational Safety and Health Act was passed in 1970. American workplaces and the workers employed there are in the midst of profound changes that will persist well into the next century. In recognition of that fact, the National Institute for Occupational Safety and Health, which plays a central role in the education and training of occupational safety and health (OSH) professionals, asked the Institute of Medicine to analyze these changes in detail, assess the supply of, demand for, and knowledge, skills, and abilities of occupational safety and health professionals, and identify personnel needs, skills, and curricula needed for the coming decades.

The committee responsible for this report found that the American workforce is becoming more diverse in age, gender, race, and nationality, and that the products of work are increasingly services rather than goods. A smaller percentage of workers are employed in large fixed industries, and a higher proportion are employed in small firms, temporary jobs, or at home. More work is now contracted, outsourced, and part time. These changes complicate implementation of workplace health and safety programs and argue for more comprehensive curricula, multidisciplinary training, and new types of training programs and delivery systems. These innovations will strengthen the traditional university-based model of four primary OSH disciplines that has guided education and training in the field to date.

The report concludes that the continuing burden of largely preventable occupational diseases and injuries and the lack of adequate OSH services in most small and many larger workplaces indicate a clear need for more OSH professionals at all levels. The authoring committee also concludes that OSH education and training needs to place a much greater emphasis on injury prevention and that current OSH professionals need easier access to more comprehensive and alternative learning experiences. In addition, provision of adequate training for the majority of American workers will depend upon the discovery of new and improved ways of reaching small and mid-sized industries with increasingly decentralized and highly mobile workforces. Ten recommendations address current and future OSH workforces and training programs.

INTRODUCTION

Each day, more than 16,000 Americans are injured on the job (over 6 million each year) and 20 more die as a result of job-related injuries (over 7,000 each year) (Bureau of Labor Statistics, 1998a). The incidence of occupational illness is more difficult to estimate, but a recent study placed the number of new cases in 1992 at 860,000 and the number of deaths as a result of occupational illnesses at 60,000 annually (Leigh et al., 1997). The economic costs of these job-related injuries ($145 billion) and illnesses ($26 billion) are much higher than those for AIDS and Alzheimer's disease and are on par with those for cancer and for circulatory diseases (Leigh et al., 1997).

The U.S. Congress passed the Occupational Safety and Health Act of 1970 to assure "every working man and woman in the United States safe and healthful working conditions." This mandate gave rise to the Occupational Safety and Health Administration (OSHA) and the National Institute for Occupational Safety and Health (NIOSH). Over the last 30 years, OSHA and NIOSH have implemented education and training programs for employers, workers, and occupational safety and health (OSH) professionals, and these programs are essential tools in reducing the burden of occupational injury and illness.

The work environment has changed considerably in the decades since the Occupational Safety and Health Act was passed. Injuries and illnesses that were unrecognized at that time now contribute significantly to the present OSH burden. The workforce has also changed, with more women, minorities, and persons with a disability in the workforce now than ever before. In addition, the numbers of workers over 50 years of age and workers under 18 years of age are also increasing. Workplaces have also evolved dramatically as a result of the U.S. economy's transition from a manufacturing base to services, and now to information and information technology. There have also been profound changes in the way in which work is organized. Distributed work arrangements, flexible matrix- and team-based organizational structures, and nonstandard work arrangements, among others, have become commonplace, challenging the traditional model for the provision of OSH programs. Important changes are also occurring in the health care system, most notably the increased emphasis on managed care and other means of reducing costs. As yet unexplored are the implications of this new care delivery system for occupational physicians, as well as possible changes in the roles of primary care physicians, nurse practitioners, and physician assistants who may be treating workers.

From a regulatory standpoint, OSHA has added job safety standards over the years that include requirements that "qualified," "designated," or "competent" persons ensure their enforcement at the work site, but there has been no agreement as to what type of training might enable such personnel to meet these requirements. OSHA also mandates training of workers in more than 100 of its standards, but it does not speak to the quantity, quality, or efficacy of that training. Few if any standards call for the training of employers or of managers responsible for workplace safety and health.

Given the widespread changes affecting nearly every aspect of workplace safety and health, NIOSH, with support from OSHA, the National Institute of Environmental Health Science (NIEHS), and the U.S. Department of Veterans Affairs, asked the Institute of Medicine to characterize and assess the current U.S. workforce and work environment and forecast the demand and need for, and supply of, qualified OSH professionals. The goal of the assessment was to identify gaps in OSH training programs that could be filled by either public or private programs and to identify the critical curricula and skills needed to meet these evolving occupational safety and health concerns.

CHARGE TO THE COMMITTEE

The charge to the committee was fourfold, calling for analyses of both the adequacy of the current OSH workforce and training and adjustments that might be required in the future because of changes in the workforce, the workplace, the organization of work, and health care delivery.

1. Assess the demand and need for OSH professionals as well as the adequacy of the OSH professional supply by sampling members of the OSH community (industry, small business, labor, academia, professional organizations, health providers, contract services, and governmental agencies). This assessment would determine the number and type of personnel currently employed; their professional duties, skills, abilities, and knowledge; and shortfalls in these categories.

2. Analyze changes in the workforce and work environment that are

affecting the roles of OSH professionals now, and how they are likely to affect these roles over the next decade.

3. Identify gaps in current OSH education and training. For example, determine whether training programs provide an appropriate number of personnel and a matrix of knowledge, skills, and abilities at appropriate levels. Determine which disciplines and skills will be most effective in addressing OSH needs over the next decade.

4. Identify the critical curricula and skills needed to meet these evolving OSH concerns.

The remainder of this summary, and the report itself, generally follow the four elements of the charge. They begin by describing OSH professionals—who they are and what they do and do not do. The following four sections then describe changes in the work environment that affect OSH and the education and training of OSH professionals. The report then describes current education and training programs and suggests changes in light of trends in the work environment.

OCCUPATIONAL SAFETY AND HEALTH PROFESSIONALS

Without a massive survey of U.S. employers, it is impossible to estimate or describe the full spectrum of OSH personnel who provide services to the U.S. workforce. However, the committee was able to describe the four traditional, or core, OSH professions—occupational safety, industrial hygiene, occupational medicine, and occupational health nursing—as well as three other disciplines that are likely to play a substantial role in the workplace of the future: ergonomists, employee-assistance professionals, and occupational health psychologists.

Although each of the four traditional OSH professions emphasizes different aspects of the field, all four share the common goal of identifying hazardous conditions, materials, and practices in the workplace and helping employers and workers eliminate or reduce the attendant risks. Occupational safety professionals, although concerned about all workplace hazards, have traditionally emphasized the prevention of traumatic injuries and fatalities. Similarly, industrial hygienists, although they do not ignore injuries, focus on the identification and control of health hazards associated with acute or chronic exposure to chemical, biological, or physical agents. Occupational medicine physicians and occupational health nurses provide clinical care and programs aimed at health promotion and protection and disease prevention. These services include not only diagnosis and treatment of work-related illness and injury, but also pre-placement, periodic, and return-to-work examinations, impairment evaluations, independent medical examinations, drug testing, disability and case management, counseling for behavioral and emotional problems that affect job performance, and health screening and surveillance programs.

Approximately 76,000 Americans are active members of the professional societies representing the core OSH disciplines. The literature suggests that as many as 50,000 more are eligible for membership by virtue of their current employment. The committee therefore estimates the current supply of OSH professionals at between 75,000 and 125,000. The committee could not find good, independent data to support an estimate of demand (i.e., the number of positions available), but the overall supply seems to be roughly consonant with employer demand. However, the committee notes that considerable need exists beyond the current demand for OSH professionals by employers. Doctoral-level safety educators are needed to teach and train injury prevention and safety professionals or their number will decrease, and both occupational medicine and occupational health nursing clearly need more specialists with formal training. Most important, a large proportion of the American workforce is outside the sphere of influence of OSH professionals, particularly those whose focus is primarily prevention, principally because few of those professionals are employed by small businesses and establishments, and, in some sectors of the economy like agriculture and construction, the workplace and the workforce are often transient.

CHANGING DEMOGRAPHICS OF THE WORKFORCE

Projected changes over the next decade will result in a workforce with a larger proportion of individuals over age 55, women, African Americans, Hispanics, and Asians. The special characteristics of these populations will need to be taken into account. For example, women and older workers have lower injury and illness rates than the labor force as a whole, although injured older workers take longer to return to work. In addition, the Americans with Disabilities Act of 1990 mandated reasonable accommodation for workers with a disabling condition, and as a result, the number of employed persons with a disability has increased sharply in the 1990s.

The committee concludes that all aspiring OSH professionals must be made aware of ethnic and cultural differences that may affect the implementation of OSH programs. In addition, the committee believes that health and safety programs are social as well as scientific endeavors and that OSH disciplines and OSH professional groups should reflect the social makeup and diversity of thought and experience of the societies they serve. Thus, the training and recruitment of OSH professionals should include all racial and ethnic groups. Further, all OSH education will need to include instruction on changes in the physical and cognitive abilities of older workers, the interaction of disabilities and chronic diseases with workplace demands, and in communication skills needed to reach minority workers, workers with low levels of literacy, and those for whom English is a second language.

CHANGING WORKPLACE

The industrial and occupational components of the U.S. economy changed significantly during the decade ending in 1998. Among goods-producing sectors, only construction added jobs, while manufacturing and mining both lost jobs. The service-producing sector, on the other hand, led by retail trade and business and health services, has grown dramatically. Four occupational groups—the executive, professional, technical, and service groups—are projected to grow more rapidly than the overall economy and, consequently, to increase their share of employment as this trend continues. Although there are some important exceptions, the rate of occupational injuries has been higher in declining industries, such as manufacturing, than in the industries that are expected to grow, such as retail trade. The majority of U.S. workers are now employed by firms with less than 100 employees; small firms showed the greatest growth in employment in the 1990s, and that trend is expected to continue. More work will be contracted, outsourced, or done on a part-time basis in the coming years. Substantial numbers of workers will hold multiple jobs, and they will change jobs more frequently. An increasing number of workers will work at home. In many sectors, the number of workers represented by unions is falling.

The committee concludes that these changes, as a whole, describe a workplace that is very different from the large fixed-site manufacturing plants in which OSH professionals have been most frequently employed. The changes complicate the implementation of OSH programs and argue for training and delivery systems that are different from those that have been relied upon to date. Simply increasing the numbers or modifying the training of OSH professionals will not be sufficient to meet these challenges, since the primary difficulty will be provision of training to either underserved workers or underserved workplaces. Extensive new regulation is possible but seems unlikely. Other problems not susceptible to site- or group-specific interventions (e.g., smoking, seat belt use, and drunk driving) have been attacked with broad public education campaigns. Future OSH professionals will need to be knowledgeable about and willing to work with mass media to reach workers at home as well as at work. The committee calls for systematic exploration of new models for implementing OSH programs for the full spectrum of American workers.

CHANGING ORGANIZATION OF WORK

Globalization, technology and other work design factors, and organizational design innovations also present training needs for OSH professionals. Increasing reliance on computer technology, distributed work arrangements, the increased pace of work, and the increased diversity of the workforce create several challenges for OSH personnel. First, new hazards could potentially emerge, both through the introduction of new technologies and through the performance of work in a more dispersed or virtual organization. Second, businesses are becoming smaller and "flatter" (i.e., fewer levels of management) and are redefining the content of work and the nature of the employment relationship. They are under pressure to compete for talent, innovate, provide exceptional quality, and bring products and services to market quickly at competitive prices. The effects of these business developments on workers include demands for new skills and continuous learning, expanded job scopes, an accelerated work pace, and the need to deal with changing workplaces. Workers also face uncertainty in employment relationships, increased interaction with both customers and coworkers, and more involvement with information and communications technologies. Further, societal developments like the increasing numbers of single parents, dual-career households, and aged dependents challenge workers to manage multiple and competing interests in their work and home lives. These factors are a major source of time conflict and carry the potential for causing dysfunction and distress in America's workforce and workplaces.

The committee concludes that OSH personnel must be well aware of the effects that these changing structural and contextual work conditions have on workers' well-being and health. They need to be able to recognize and react to effects of these work organization factors on cognitive and behavioral functioning, including stress-related conditions and their link to health, safety, and performance. Finally, OSH personnel need to have a basic competence in prevention and intervention strategies.

CHANGING DELIVERY OF HEALTH CARE

Physicians and nurses specializing in occupational health, and the institutions within which they work, must operate as part of a health care system that has been undergoing profound changes during the 1990s. One of the most striking features of health care reform has been the dramatic growth of managed care, a major element of which is tighter control on the utilization of health services. This has led to an emphasis on caregivers seeing more patients (vs. providing preventive services, receiving and providing additional education, and performing research), the increased use of primary care physicians and paraprofessionals instead of specialty-trained physicians, and a consolidation of small practices and clinics into large occupational health clinics and integrated systems providing full-time coverage (24 hours a day, 7 days a week) of workers and their families.

> *The committee concludes that all health care professionals need to be more familiar with workers' compensation law, and that aspiring OSH professionals need training in the principles of health care organization and financing, managed care, and multi-disciplinary health care.*

EDUCATION AND TRAINING PROGRAMS

Any consideration of the future OSH workforce must include an analysis of the educational "pipeline" from which these professionals emerge. The committee used a variety of sources to assemble estimates of the annual number of master's-level graduates in the four core OSH disciplines. Twenty-nine U.S. schools offer such degrees in occupational safety, and they graduate about 300 students annually. This number is extremely low, given the incidence of workplace injuries, but employers' apparent willingness to hire graduates with baccalaureate degrees in occupational safety (about 600 annually) limits the demand for master's-level safety professionals. Less than 10 students are awarded doctoral degrees in occupational safety each year, a level low enough to threaten the future viability of academic departments of occupational safety. The committee estimates that approximately 400 master's-level industrial hygienists graduate each year, a volume roughly equal to employer demand in the industrial sector that has most commonly used them. Forty institutions offer occupational medicine residencies, and they produce about 90 graduates annually, a number that is insufficient to replace existing occupational medicine specialists when they leave practice. Attracting applicants to this field is a major problem, since the field draws a large proportion of its practitioners from among established physicians, for whom a return to full-time student status is not feasible. A similar situation exists in nursing. Twenty-one schools of nursing and public health award only about 50 master's-level degrees in occupational health nursing each year. Curricula in all four OSH disciplines are predominantly technical and science based, with an engineering and physical science emphasis in safety and industrial hygiene and a biological, health, and programmatic emphasis in nursing and medicine. NIOSH training programs provide grants totaling approximately $10 million per year in support of OSH professional education, resulting in approximately 500 OSH-related degrees (or completed residencies) annually. Occupational medicine receives the most funding, reflecting the high cost of postgraduate specialist training for licensed physicians. Industrial hygiene is a close second, with occupational health nursing (at about 55% of the level of occupational medicine's funding), and occupational safety (at about 33% of the level of occupational medicine's funding), receiving considerably less support. The committee also reviewed worker and manager training provided by OSHA and others. No degrees are associated with this training, which takes many forms, from simple handouts and videotape viewings to 1–2 weeks of classroom and hands-on instruction. The committee did not attempt an exhaustive survey, but it is clear that tens of thousands of hours of worker training are done, largely in response to OSHA mandates.

The committee concludes that current problems in the education and training of OSH professionals include an insufficient emphasis on the prevention of traumatic injury, which is reflected most clearly in the very small number of doctoral-level graduates in occupational safety; the limited support for relevant research in departments other than those that grant OSH degrees; and an inability to attract physicians and nurses to formal academic training in OSH. An existing problem likely to be exacerbated by the many changes under way in the work environment is the narrow focus of education and training programs on OSH personnel who have traditionally served large, fixed-site manufacturing industries. A problem in terms of responding to changes in the future workplace is a lack of research and training in a number of areas of increasing importance: behavioral health, work organization, communication, management, team learning, workforce diversity, information systems, prevention interventions, and evaluation methods, among others. Additional topics in need of attention include methods for effective training of adult workers; the physical and psychological vulnerabilities of members of the workforce stratified by age, gender, and socioeconomic, and cultural background; the resources available to help with injury prevention and analysis; business economics and values; health promotion and disease prevention; community and environmental concerns; and the ethical implications of technological advances such as the mapping of the human genome.

*The committee also concludes that health and safety training for work-
ers, although abundant, is of unknown quality and efficacy, and that
OSH training for managers is rarely demanded, offered, or requested.*

ALTERNATIVES TO TRADITIONAL
EDUCATION AND TRAINING

A substantial portion of the current OSH professional workforce consists of people who do not have advanced degrees or, in some cases, even
baccalaureate degrees. These people were attracted to OSH well after
beginning their working lives, when full-time attendance in traditional
education programs was not a viable option. Now, however, techniques
such as distance education and alternative training programs allow students to continue their education outside the confines of a traditional
classroom. Distance education is a planned and structured means of learning that uses electronic technology involving audio, print, video, and
Internet media, alone or in combination. Limited but impressive data on
the popularity and effectiveness of distance education in preparing physicians for occupational medicine board-certification examinations point to
its potential as a means of facilitating education and certification of the
many practicing OSH personnel without formal specialty training in the
area.

*The committee concludes that although traditional approaches remain
indispensable for some types of instruction, NIOSH should develop incentives to promote the use of distance education and other nontraditional approaches to OSH education and training. An integral part of
these innovative programs should be a thorough evaluation of both program content and the performance of their graduates on, for example,
credentialing examinations and in job placement compared to that of
graduates of traditional programs.*

SUMMARY OF FINDINGS AND RECOMMENDATIONS

The charge to the committee called for analyses of both the adequacy
of the current OSH workforce and training and adjustments that might be
required in the future because of changes in the workforce, the workplace, the organization of work, and health care delivery.

Current OSH Workforce and Training

The current supply of OSH professionals, though diverse in knowledge and experience, generally meets the demands of large and some

medium sized workplaces. However, the burden of largely preventable occupational diseases and injuries and the lack of adequate OSH services in most small and many medium-sized workplaces indicate a need for more OSH professionals at all levels. The committee also finds that OSH education and training should place more emphasis on injury prevention and that current OSH professionals need easier access to more comprehensive and alternative learning experiences.

RECOMMENDATIONS:

To address the critical need to mitigate the enormous and continuing impacts of acute and chronic injuries on worker function, health, and well-being, to develop new leaders in this neglected field, and to strengthen research and training in it at all levels:

Recommendation 1: Add a new training initiative focused on prevention of occupational injuries.

NIOSH should develop a new training initiative focused on the prevention of occupational injuries, with special attention to the development of graduate-level faculty to teach and conduct research in this area. Possible approaches would include regional Occupational Injury Research, Prevention, and Control Centers as an entirely new program or by modification of the existing NIOSH training programs or collaboration with the Centers for Disease Control and Prevention's National Center for Injury Prevention and Control.

To enhance needed multidisciplinary research in injury prevention and in occupational safety and health in general:

Recommendation 2: Extend existing training programs to support of individual Ph.D. candidates.

NIOSH should extend existing training programs to support individual Ph.D. candidates whose research is deemed of importance to the prevention and treatment of occupational injuries and illnesses, independent of academic department or program. Restricting support to students in Education and Research Centers or Training Project Grants–affiliated departments or disciplines deprives the OSH field of individuals who may have innovative responses to changing circumstances.

To address the lack of formal training among OSH professionals:

Recommendation 3: Encourage distance learning and other alternatives to traditional education and training programs.

NIOSH should encourage the use and evaluation of distance education and other nontraditional approaches to OSH education and training, especially as a means of facilitating education and certification of the many practicing OSH personnel without formal specialty training in the area.

Recommendation 4: Re-examine current pathways to certification in occupational medicine.

The American Board of Preventive Medicine should reexamine the current pathways to certification in occupational medicine. Specifically, it should consider

• extending eligibility for its existing equivalency pathway to include physicians who graduated after 1984 and
• developing a certificate of special competency in occupational medicine for physicians who are board certified in other specialties but who have completed some advanced training in occupational medicine.

Future OSH Workforce and Training

Expected changes in the workforce and in the nature and organization of work in the coming years will result in workplaces that will be quite different from the large fixed-site manufacturing plants in which OSH professionals have previously made their greatest contributions. The delivery of OSH services will become more complicated, and additional types of OSH personnel and different types of training than have been relied upon to date will be needed. Simply increasing the numbers or modifying the training of occupational health professionals will not be sufficient, since the primary difficulty will be to provide training to underserved workers and underserved workplaces. Traditional OSH programs must be supplemented by a new model that focuses on these workers and work sites.

RECOMMENDATIONS:

To help ensure high-quality occupational safety and health programs for the full spectrum of American workers:

Recommendation 5: Solicit large-scale demonstration projects that target training in small and mid-sized workplaces.

NIOSH, in collaboration with OSHA, should fund and evaluate large-scale demonstration projects that target training in small and midsized workplaces. These innovative training programs should encourage the use of new learning technologies, should include a recommended core of competencies, and could lead to the creation of a new category of health and safety personnel—OSH managers.

Recommendation 6: Evaluate current worker training and establish minimum quality standards.

OSHA should join together with NIOSH, NIEHS, unions, industries, and employer associations to evaluate the efficacy of OSHA and other worker training programs and better define minimum training requirements.

Recommendation 7: Solicit demonstration projects to create model worker training programs for occupational safety and health trainers.

NIOSH, in collaboration with OSHA, should fund demonstration project grants that target specific employment sectors as an incentive to develop model training programs for another category of health and safety personnel—OSH trainers.

To address the challenges posed by the increasing diversity of the U.S. workforce:

Recommendation 8: Increase attention to special needs of older, female, and ethnic/cultural minority workers.

All aspiring OSH professionals must be made aware of ethnic and cultural differences that may affect implementation of OSH programs. In addition, because OSH programs are social as well as scientific endeavors, NIOSH, OSHA, NIEHS, other federal and state agencies, educational institutions, unions, employers, associations, and others engaged in the training of OSH personnel should foster and/or support efforts to provide a body of safety and health professionals and trainees that reflects age, gender, and ethnic/cultural background of the workforces that they serve. These organizations should also foster meaningful instruction on the aging process, the interaction of disabilities and chronic diseases with workplace demands, and communication skills to interact with minority and workers with low levels of literacy and those for whom English is a second language.

To prepare present and future OSH professionals to address continuing changes in the U.S. workforce, in the workplace, and in the organization of work itself as major determinants of workplace safety, health, and well-being:

Recommendation 9: Examine current accreditation criteria and standards.

Boards and other groups that accredit academic programs in the OSH professions, in conjunction with appropriate professional organizations, should carefully examine their current accreditation criteria and standards, paying special attention to the needs of students in the areas of behavioral health, work organization, communication (especially risk communication), management, team learning, workforce diversity, information systems, prevention interventions, healthcare delivery, and evaluation methods.

Recommendation 10: Broaden graduate training support to include behavioral health science programs.

NIOSH should broaden its graduate training support to include the behavioral health sciences (e.g., psychology, psychiatry, and social work) by developing and maintaining training programs in work organization and the prevention and treatment of physical and mental effects of work-related stress.

BOX 1 SUMMARY OF RECOMMENDATIONS

Current OSH Workforce and Training

1. Add a new training initiative focused on prevention of occupational injuries.
2. Extend existing training programs to support of individual Ph.D. candidates.
3. Encourage distance learning and other alternatives to traditional education and training programs.
4. Re-examine current pathways to certification in occupational medicine.

Future OSH Workforce and Training

5. Solicit large-scale demonstration projects that target training in small and mid-sized workplaces.
6. Evaluate current worker training and establish minimum quality standards.
7. Solicit demonstration projects to create model worker training programs for occupational safety and health trainers.
8. Increase attention to special needs of older, female, and ethnic/cultural minority workers.
9. Examine current accreditation criteria and standards.
10. Broaden graduate training support to include behavioral health science programs.

1

Introduction

Although improvement in workplace safety was recently hailed as one of the 20th century's outstanding achievements in public health (Centers for Disease Control and Prevention, 1999) (Figure 1-1), it is nevertheless true that even in the waning years of the century, 20 American workers died each day as a result of occupational injuries, for a total of 6,000 deaths per year. More than 16,000 suffer nonfatal injuries on the job every day, for a total of 6 million injuries per year (Bureau of Labor Statistics, 1998a). The incidence of occupational illness is more difficult to estimate, but a recent thorough and methodologically sound attempt placed the number of new cases in 1992 at 860,000 and the number of deaths from occupational disease that year alone at 60,000 (Leigh et al., 1997). The best current estimate of the costs of these job-related injuries ($145 billion) and illnesses ($26 billion) is much higher than those for AIDS and Alzheimer's disease and are on par with the costs of cancer or of all circulatory diseases (Leigh et al., 1997).

The U.S. Congress passed the Occupational Safety and Health Act of 1970 (OSHAct) to assure "every working man and woman in the United States safe and healthful working conditions." This mandate gave rise to the Occupational Safety and Health Administration (OSHA) and also established the National Institute for Occupational Safety and Health (NIOSH). Over the last 30 years NIOSH has implemented a training program to address a provision of the OSHAct that mandates an "adequate supply of qualified professionals to carry out the purposes of the Act." NIOSH extramural funds support a $10 to $15 million training program

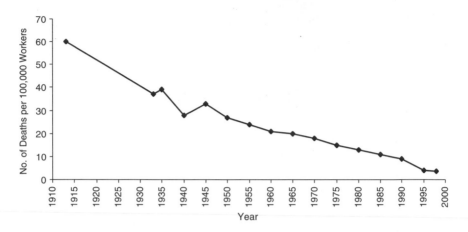

FIGURE 1-1 Deaths per 100,000 workers due to injury, 1913 to 1998. (Data for death rates from injury for the years between 1913 and 1933 were not available. Because of adoption of the Census of Fatal Occupational Injuries to include deaths of any worker regardless of age (previous years surveyed only persons over the age of 16 who were gainfully employed), numbers of deaths and death rates from 1992 to the present are not comparable to those for earlier years). SOURCES: Corn (1992), National Safety Council (1999).

(approximately 9 percent of the institute's occupational safety and health [OSH] budget) that comprises a network of 15 regional Education and Research Centers (ERCs) located at leading universities in 14 states and approximately 40 individual training project grants in 25 states and Puerto Rico. Every year, approximately 500 students graduate from NIOSH-supported programs with backgrounds in occupational medicine, occupational health nursing, industrial hygiene, and safety. Many current leaders and OSH advocates have received training from NIOSH-funded programs. According to a 1996 report by the inspector general of the U.S. Department of Health and Human Services (Office of Inspector General, 1996), about 90 percent of ERC graduates pursue careers in OSH in private industry, labor, government agencies, medicine, and academia.

   The work environment has changed since the OSHAct was passed more than 30 years ago. Work injuries and illnesses that were unrecognized at that time now contribute greatly to the present OSH burden. New information about ergonomics and the association between work organization and job stress mean that there is a newly recognized need for personnel qualified in carrying out interventions that will prevent adverse health outcomes related to these factors. Worker demographics are changing, with more women, minorities, and people with disabilities being in the workforce. Older workers (those over 50 years of age) and younger workers (those under 18 years of age) also play important roles in the workforce. Individuals in these populations have risk factors that may need to be addressed differently from those risk factors among individuals in the general population. For example, are older workers more adversely affected by shiftwork? The U.S. workplace itself has also evolved as the U.S. economy has moved rapidly from a manufacturing-based to a service-based economy and is now developing into one centered on information and information technology. In addition, there have been profound changes in the way in which work is organized. Distributed work arrangements, flexible matrix- and team-based organizational structures, and nonstandard work arrangements have become commonplace, challenging the competencies of current occupational safety and health professionals.

In terms of delivery of OSH services, industry-supplied OSH resources may no longer play as large a role as they once did. As companies downsize, OSH staff may be outsourced or eliminated altogether. The OSH role may be assumed by other staff who do not have training in OSH or by a person with training in a combination of disciplines. Other important changes are occurring in health care delivery generally, with an increased emphasis on managed care and other means of reducing costs. The evolving role of the occupational physician has not been explored in this new delivery system, nor have the roles of primary care physicians, nurse practitioners, or other health professionals who may be treating workers.

From a regulatory standpoint, OSHA has added standards over the years that require "qualified," "designated," or "competent" persons to ensure enforcement at the work site, but there has been no agreement on the training that will enable personnel to meet these designations. OSHA also mandates training of workers in more than 100 standards but does not speak to the quantity, quality, or efficacy of that training. Few if any standards call for training of employers or managers responsible for workplace safety and health.

Because of these workforce, workplace, and OSH care changes, NIOSH's traditional focus on training of industrial hygienists, occupational physicians, occupational nurses, and safety professionals may no longer be sufficient. Evolving demands on the OSH professional in meeting the challenges of the new workplace may call for a much broader perspective than that taken in the past, including an expanded emphasis on such fields as epidemiology, ergonomics, health communication, the behavioral sciences, health care cost control, and management.

There is thus an urgent need to examine the numerous factors changing the modern workplace, derive the implications of these trends for OSH, and make corresponding changes in the education and training of

OSH professionals, management and union officials, and the workers themselves. Moreover, national guidelines are required to inform universities, industries, and other public health agencies of the personnel requirements needed to ensure worker safety and health. Accordingly, NIOSH, with support from OSHA, the National Institute of Environmental Health Science (NIEHS), and the U.S. Department of Veterans Affairs, approached the Institute of Medicine (IOM) for a needs assessment that would characterize the current U.S. workforce and work environment and forecast the demand, need, and supply of qualified OSH professionals.* The goal of the assessment would be to identify gaps in occupational safety and health training programs in the United States that can be filled by either public or private programs and identify the critical curricula and skills needed to meet these evolving OSH concerns.

CHARGE TO THE COMMITTEE

The charge to the committee from the sponsors was fourfold, calling for analyses of both the adequacy of the current OSH workforce and training and adjustments that might be required in the future because of changes in the workforce, the workplace, the organization of work, and health care delivery.

1. Assess the demand and need for OSH professionals as well as the adequacy of the OSH professional supply by sampling members of the OSH community (industry, small business, labor, academia, professional organizations, health providers, contract services, and governmental agencies). This assessment would determine the number and type of personnel currently employed; their professional duties, skills, abilities, and knowledge; and shortfalls in these categories.

2. Analyze changes in the workforce and work environment that are affecting the roles of OSH professionals now, and how they are likely to affect these roles over the next decade.

3. Identify gaps in current OSH education and training. For example, determine whether training programs provide an appropriate number of personnel and a matrix of knowledge, skills, and abilities at appropriate levels. Determine which disciplines and skills will be most effective in addressing OSH needs over the next decade.

4. Identify the critical curricula and skills needed to meet these evolving OSH concerns.

---

*Also contributing to the support of the project were the American Association of Occupational Health Nurses and the American Academy of Industrial Hygiene.

LEGISLATIVE AND REGULATORY BACKGROUND

Laws and regulations have played a major role in the great improvement in worker health and safety alluded to in the opening lines of this report. Appendix C provides the names and dates of some of the most significant events in the history of OSH. The process has been a cumulative one, however, so this section will summarize the major provisions of the most recent and far-reaching of these laws, the OSHAct of 1970 (the Mine Occupational Safety and Health Act, as amended in 1977, is very similar but is limited to a single industry) and the common features of state workers' compensation (WC) laws. Workers, employers, and occupational safety and health professionals are affected by many more industry-specific laws and regulations, but these two affect all of these individuals and are thus the starting place for OSH education and training.

The Occupational Safety and Health Act of 1970

The OSHAct was a far-reaching piece of legislation that went far beyond spawning the NIOSH training programs mentioned above. It authorizes the Secretary of Labor to set mandatory OSH standards for businesses, conduct inspections to see that they are being observed, and apply sanctions for violations of the standards. Coverage extends to all businesses except the self-employed, farms that employ only members of the immediate family, and a few specific industries such as mining, nuclear power, and civil aviation that are regulated by other federal agencies. It does not cover over 8 million employees of federal, state, and local governments, including public schools, public healthcare facilities, and correctional institutions. Federal agencies, although not generally subject to inspection by OSHA, are required to comply with standards consistent with those issued to the private sector, and they must self-inspect annually. Like private-sector employers, they must also record and analyze injury and illness data and provide training to protect employees from on-the-job hazards. OSHA is the federal agency responsible for promulgating and enforcing the standards that employers must meet, but the OSHAct does allow states to assume that role within their own borders, and 25 states have done so to date. States seeking authority to establish their own OSH programs must submit a detailed plan to OSHA for approval, which depends upon convincing OSHA that the state program will be at least as effective as the federal program (states must add or modify their standards whenever OSHA adds or modifies the federal standards). Once approved, OSHA funds up to 50 percent of the state program. Further use of the term "OSHA" in this chapter should be taken to mean OSHA or OSHA-approved state programs.

Although the process of issuing a new standard is long and cumbersome, to ensure that all interested parties can be heard, there are hundreds of standards covering a wide range of hazardous materials, equipment, procedures, and working conditions. These standards require employers to protect employees in a variety of ways: from design and maintenance of the workplace and equipment to training of the worker to perform his or her job safely. General Industry Standards apply to materials and conditions common to many industries (for example, toxic chemicals, fire protection, electrical safety, material handling, and general environmental controls). Other standards may be highly specific applications to limited aspects of particular industries (scaffolding, fall protection, or surface transportation of explosives or bloodborne pathogens, for example).

OSHA is authorized to enter and inspect workplaces to enforce these standards. Such inspections cover both record-keeping (employers are required to keep standardized records of injuries and illnesses, and many standards require documentation of exposure assessments, medical surveillance, training, and other activities relevant to OSH) and firsthand examination of the work site, including consultation with employees.

OSHA and its 25 state partners have had some undeniable successes. Since 1970, the overall workplace injury death rate has been cut in half. OSHA's cotton dust standard virtually eliminated brown lung disease in the textile industry; and OSHA's lead standard reduced lead poisoning in battery plant and smelter workers by two-thirds. OSHA inspections may be an important part of that success: according to a recent study, in the 3 years following an OSHA inspection that results in penalties, injuries and illnesses drop on average by an average of 22 percent. Overall injury and illness rates have declined in the industries where OSHA has concentrated its attention yet have remained unchanged or have actually increased in the industries where OSHA has had less of a presence. The last of those observations assumes real importance when one notes that OSHA resources allow inspection of less than 2 percent of U.S. workplaces every year. The Mine Safety and Health Administration, in contrast, inspects each U.S. mine at least four times yearly (Institute of Medicine, 1998).

The OSHAct went beyond regulation and enforcement by also charging the Secretary of Labor with establishing and supervising education and training programs for employers and employees. These programs covered recognition and prevention of unsafe or unhealthful working conditions. In addition, the Secretary of Health, Education, and Welfare (now the Secretary of Health and Human Services) was charged with conducting "educational programs to provide an adequate supply of qualified personnel to carry out the purposes of this [OSH] Act" and "informational programs on the importance and proper use of safety and health equipment." OSHA has assumed the lead in OSH training of workers and managers, and NIOSH has focused on degree programs and continuing education in occupational medicine, occupational health nursing, industrial hygiene, and occupational safety. Although its effectiveness has been questioned (for example, Hale [1984] and Tan et al. [1991]), health and safety training of workers has generally been acknowledged to be an important component in prevention of work-related injuries and illnesses (Office of Technology Assessment, 1985: Keyserling, 1995; Cohen and Colligan, 1998). Both agencies' programs are described in more detail in Chapter 7.

Workers' Compensation

Workers' Compensation (WC) is no-fault insurance for occupational injuries and illnesses. Beginning in the first decade of the 20th century, each of the 50 states and the District of Columbia have passed "workers' comp" laws that vary widely in coverage and benefits but that, in general, provide employers immunity from lawsuits in exchange for payments to the affected worker or dependent survivor. Immunity can be waived in cases in which the employer is not in compliance with relevant OSHA standards (in practice, it usually takes gross negligence or criminal activity on the part of the employer), so the laws serve as an incentive for employers to comply with OSHA standards. In addition, since premiums are determined in part by the numbers and sizes of claims, employers should have a financial interest in safe working conditions.

Claims arising from work-related injuries often stem from obvious one-time events where neither the existence of the injury nor the fact that it is work-related is in doubt (a whaler gets his foot tangled in a harpoon line, is dragged into the water and drowns). In other cases, one or both of these criteria (an injury has occurred, and it occurred as a result of employment) are difficult to establish. For example, low back pain is a very common complaint that often cannot be substantiated with objective medical evidence, and injuries like carpal tunnel syndrome are characterized by a gradual onset that may leave room for doubt about whether a specific case is truly occupational in nature. Occupational illnesses can be even more difficult to assess, and only a fraction of cases of work-related disease are thought to be covered by workers' compensation. Occupational diseases are seldom the result of a single identifiable incident, and many, like cancer, may be diagnosed only many years after exposure to the causative agent. A solution adopted in many states was to establish a list or schedule of recognized occupational diseases associated with certain trades, materials, or processes. This has generally been abandoned as too limiting, and replaced by a case-specific judgment on whether the illness arose out of the activities of work.

The benefits paid include compensation not only for the explicit costs of medical treatment, including rehabilitation if necessary, but also for the loss of income incurred, which on average is about 50 to 60 percent of the total benefit. Funeral and burial costs are covered in fatal cases, and there is generally some schedule of payment for permanent disabilities independent of any attendant loss of income. The National Academy of Social Insurance (Mont et al., 1999) estimates that approximately $40 billion in benefits was paid out annually in 1993 through 1996, the last year for which they have data, although other estimates are as much as $70 billion to $100 billion (Leone and O'Hara, 1998). A number of studies suggest that WC is underutilized, however, with as much as 30 to 60 percent of work-related fatalities not found in WC records (Cone et al., 1991; Stout and Bell, 1991; Leigh et al., 1997). Parker et al. (1994) found that 67 percent of eligible injuries were not reported to the WC system. Reasons proposed for this underutilization include workers' ignorance of their rights, workers' desire to be "team players," and fear of retaliation from the employer, who has a financial stake in minimizing claims.

The Americans with Disabilities Act

The Americans with Disabilities Act (ADA) was signed into law on July 26, 1990, with the intent of protecting from discrimination the millions of disabled Americans in the workforce or those who are seeking to enter the workforce. A disabled person is defined as an individual who has a physical or mental impairment that substantially limits one or more of his or her major life activities, has a record of such an impairment, or is regarded as having such an impairment. ADA consists of five different titles, the most important being Title I, which prohibits employing entities from discriminating against a qualified, disabled individual in any aspect of employment. Titles II, III, and V of ADA address the need for construction to allow disabled individuals access to public areas and services, as well as to private businesses and recreational establishments (Rogers, 1994). Title IV mandates that various forms of telecommunications services be offered to those individuals with hearing impairments. The U.S. Department of Justice (DOJ) has been given the authority to issue regulations and to offer expertise, technical assistance, and enforcement for Titles II and III of ADA. DOJ is also responsible for calibrating state and municipal codes for building construction and building alterations so that they adhere to ADA standards (U.S. Department of Justice, 1999).

As a result of the new responsibilities associated with adherence to ADA, OSH personnel have assumed a greater presence and new roles in the workplace (Rogers, 1994). For example, OSH personnel are increasingly important to the recruitment process. To avoid inadvertent discrimination against prospective employees, employing entities must provide accurate job descriptions, which should include input from OSH personnel. Furthermore, when employing entities assess candidates for a new position, the assessment of the position by an OSH expert may be one of the factors that determines the prospective employee's fitness to complete essential job functions. In assessing a job, OSH personnel consider the general work environment, tools associated with the position, stressors, and other elements that affect essential job functions.

Other roles that OSH personnel play as a result of ADA lie in the areas of reasonable accommodation and WC. Reasonable accommodation provides those disabled persons who are qualified for a particular position with the means to overcome workplace barriers, be they physical, communicative, scheduling, or simple prejudice. Occupational safety personnel and ergonomists, by providing suggestions for reasonable accommodations, become important figures in helping disabled individuals and the employing organizations in overcoming physical barriers. In regard to financial compensation, medical examinations by occupational medicine physicians and occupational health nurses aid in determining not only whether the cause of an injury was work related but also whether an employee has been temporarily or permanently disabled and when and in what capacity he or she is fit to return to work. As a result of ADA, OSH personnel have increased their presence in the workplace, particularly in the areas of recruitment, reasonable accommodation, and financial compensation.

PREVIOUS STUDIES OF THE OSH WORKFORCE

Several previous studies have attempted to estimate the size and nature of the OSH workforce or some part thereof and of its adequacy for the task of minimizing occupational injuries and illnesses. The largest and oldest of these is a nationwide survey of more than 3,000 nonagricultural organizations and firms and all 112 known educational institutions that provide a degree or certificate in an OSH-related field (National Institute for Occupational Safety and Health, 1978). A smaller survey study commissioned by NIOSH in 1985 (Cox and Johnston, 1985) solicited descriptive information and summary data on graduates from all identifiable OSH academic programs at U.S. colleges and universities. Additional components of the Cox and Johnston (1985) study were telephone interviews with 40 employers of OSH professionals and questionnaire data from 500 recent OSH graduates. In 1987, IOM convened a committee that identified a number of barriers that prevent primary care physicians from adequately diagnosing and treating occupationally and environmentally related diseases and published the committee's report the following year (Institute of Medicine, 1988). Follow-up reports by Castorina and Rosenstock (1990) and IOM (1991) identified a shortage of occupational and environmental physicians and offered specific measures to remedy the shortfall.

The 1978 Nationwide Survey of the OSH Workforce

The relevant universe of workplaces for the 1978 survey study's estimates of the 1977 OSH workforce included all firms that employ 100 or more individuals in mining, construction, manufacturing and transportation, communication and utilities and all firms that employ 500 or more individuals in the trade and service industries and in non-OSH related state and local government agencies. Survey instruments collected data on each firm and its OSH operations, on each of the firm's OSH-related positions, and on the background and activities of each of the firm's identified OSH personnel. A total of 3,300 firms responded representing a return rate of approximately 50 percent.

On the basis of the data from the survey, the study's authors estimated that approximately 85,000 employees spent at least 50 percent of their time in OSH-related activities. However, only about two-thirds of their positions were full-time. Manufacturing employed the greatest number of OSH employees, accounting for 44 percent of the total number of OSH employees, followed by trades, services, non-OSH related government organizations, and loss-prevention insurance carriers. Almost half of the OSH workforce (44 percent) was employed in a safety-related position (injury prevention). Nurses comprised an additional 11 percent, physicians comprised 1 percent, industrial hygienists and fire protection experts each comprised 6 percent, and radiation safety specialists comprised 2 percent of the identified OSH workforce. Twenty-four percent of the OSH workforce fell in the "general" category because although they spent 50 percent of their time on OSH-related activities, their work was spread over two or more basic areas. Overall, about 58 percent of OSH employees' time was spent on the prevention and treatment of injuries. Twenty-two percent of their time was spent on the prevention and treatment of illnesses, and about 20 percent of their time was spent on fire prevention. The OSH employees themselves reported a mean age of 42, an average of about 7 years of tenure in their current job, and just short of 15 years of formal education (i.e., less than a baccalaureate degree). Employers, asked about their expectations for new hires, anticipate only 4 years experience on average and were looking for bachelor's degrees for only one-third of their new hires and graduate degrees for only 3 percent of their new hires (1 percent would presumably be physicians).

The study's second major product was a set of predictions about demand for OSH employees in 1980, 1985, and 1990. These were produced by a model based on industry employment projections by the Bureau of Labor Statistics. One of the two published predictions assumed that the distribution and concentration of OSH employees would remain unchanged (status quo model), so the number of OSH employees would grow at the same rate as the rate of growth of the industries that employed them in 1977. A second prediction assumed that government regulations would lead to an increased concentration of OSH employees and used a panel of technical advisers to make estimates about the industries and specialties most likely to be affected (accelerated model). The two forecasts for the 1990 OSH workforce were for 104,000 in the status quo model and 110,000 in the accelerated model. This translates to a demand for roughly 3,000 new hires annually in the late 1980s.

The final segment of the report looked at the expected supply of OSH personnel from four sources: educational institutions, insurance carriers, government, and industry. A total of 956 OSH related degrees (including 208 2-year associate degrees) were awarded by the 88 institutions responding to the NIOSH request for data. This number was up sharply from the 1970 total of 144, and the responding schools anticipated a continuing increase to a level of 2,200 by 1990. Inquiries to loss prevention insurance carriers indicated that 50 percent of all new hires were trained internally—300 to 500 per year. The only federal government source included was the National Mine Health and Safety Academy, which in 1978 graduated 200 to 250 individuals annually (the OSHA Training Institute and other federal training programs were considered primarily continuing education and not primary sources of supply). It proved impossible for the report's authors to estimate the extent of formal and informal training being carried out by private firms, but they speculated that reliance on in-house training had diminished steadily since the passage of the OSHAct of 1970 and would continue to do so. Their total projected annual supplies for 1977 and 1990 were 1,426 and 2,504, respectively. The report concluded that demand equaled supply for industrial hygienists but that demand exceeded supply for safety personnel, occupational health nurses, occupational health physicians, and OSH generalists.

The 1985 Cox and Johnston Report on the Impact of OSH Training and Education Programs

NIOSH commissioned a study on the impact of OSH training and education programs (Cox and Johnston, 1985) to evaluate the effectiveness of the training programs that it had instituted to alleviate perceived shortages in trained OSH personnel. It sought to determine the nature and number of academic programs in OSH, estimate the current and projected numbers of graduates from those programs, and obtain feedback from both recent graduates and a small sample of employers about the value of the academic preparation.

The study identified 241 academic programs in OSH, at 136 U.S. institutions. NIOSH-supported ERCs accounted for 83 of the 241 programs, and another 34 programs received non-ERC-related support from NIOSH. Programs that combined industrial hygiene and safety were most common ($n = 75$), followed by industrial hygiene programs ($n = 73$) and safety programs ($n = 62$), occupational medicine programs ($n = 16$) and occupational health nursing programs ($n = 15$). Thirty-five programs offered associate degrees or certificates (20 in industrial hygiene and safety and 12 in safety); 71 programs offered baccalaureate degrees; 87 offered master's degrees (35 in industrial hygiene, 19 in safety, 20 in industrial hygiene and safety, and 13 in occupational health nursing). Doctoral degrees were offered by 33 programs (18 in industrial hygiene), and the 15 occupational medicine programs were post doctoral.

In 1979-1980 and again in 1981-1982, these programs produced about 1,600 graduates, with about 950 at the associate or baccalaureate level. Almost half of the graduate-level degrees were granted by institutions associated with ERCs. A sample of 500 ERC graduates reported ease in gaining suitable employment, a high level of job satisfaction, and a reasonable match between academic preparation and job requirements. Employers ($n = 40$) reported some shortage of occupational health nurses and occupational medicine graduates, a reasonable balance of supply and demand for industrial hygienists, and a possible surplus of occupational safety graduates.

The 1988 IOM Report on the Role of the Primary Care Physician in Occupational and Environmental Medicine

Growing recognition of adverse health effects associated with exposure to toxic substances at home, at work, and in the general community, and the limited capacity of most physicians to deal with those effects, led IOM to form a committee to seek ways to foster more active and effective participation of primary care physicians in preventing and treating occupationally and environmentally related health problems. The committee recommended improved information sources, preferably a single access point for pertinent clinical information, improved availability of clinical consultation, primarily through increased numbers of trained specialists, and changes in the reimbursement system both to increase the adequacy and speed of payment for treatment and also to reward prevention efforts (Institute of Medicine, 1988). IOM further recommended that the field of occupational and environmental medicine (OEM) be better represented on medical school faculties and in medical school curricula, that residency programs for internal medicine and family practice devote more time to experience in OEM, and that these areas receive greater representation in the examinations of the various state and local boards for certification and licensing.

The 1991 IOM Report Addressing the Physician Shortage in Occupational and Environmental Medicine

A second IOM committee commissioned a quantitative estimate of the need for and supply of physicians with clinical training in OEM, reiterated the recommendations of the 1988 study, and elaborated on the methods by which those recommendations might be implemented. The needs estimate is fully described in a paper by Castorina and Rosenstock (1990). In that critical review of previously published estimates of the need for and supply of physicians with clinical training in OEM, the authors call for 1 to 3 OEM specialists at each of the 127 U.S. medical schools, 1 to 1.5 community-based OEM specialists per 100,000 population, and 1 to 3 board-certified OEM physicians in each of 57 state and 505 local public health agencies. They also assert that approximately 1 percent of primary care physicians should have special competence in OEM. Their resulting estimate of a need for 3,100 to 4,700 board-certified OEM specialists and 1,500–2,000 primary care physicians with special competence in OEM is far higher than their estimate of the supply (1,200 to 1,500).

Central to that IOM committee's suggestions for addressing this shortage was the vision of three levels of specialization. OEM specialists, who would be board certified in occupational medicine, would be employed as full-time university faculty, in a public health agency, in industry, or as a consultant to any of those institutions. OEM clinicians, who would be board certified in internal medicine or family practice but with a Certificate of Added Qualifications in OEM, would work primarily in community hospitals and clinics as clinical faculty and in industry. Primary care physicians, who would be board certified in internal medicine or family practice but who would have additional awareness of OEM issues via implementation of the initiatives outlined in the 1988 IOM report, would continue to be the first contact with the medical system for most patients.

Not only did the committee offer the suggestion of a Certificate of Added Qualifications to create OEM Clinicians but it also proposed an alternative, streamlined approach to certification of full-fledged OEM specialists that would shorten training by a year.

METHODS OF THE PRESENT IOM STUDY

In 1999 IOM assembled a committee of scientists and medical practitioners in accordance with the established procedures of the National Academies, including an examination of possible biases and conflicts of interest and provision of opportunity for public comment. A roster with brief biographies of the committee members is provided in Appendix A.

A wide variety of sources were used to assemble the data and information necessary to respond to the charge. A list of some of the individuals who assisted the committee in this effort is provided on page *xiii*. An initial organizational and data-gathering meeting of the committee in March 1999 provided an overview of several important organizations and education and training programs within the federal government— NIOSH, OSHA, NIEHS, and U.S. Department of Veterans Affairs. At a subsequent meeting, in May 1999, the committee heard about the training and utilization of OSH professionals in additional federal agencies, namely the U.S. Department of Defense and the U.S. Department of Energy. That meeting also featured briefings from representatives of the major OSH professional associations: the American Industrial Hygiene Association (AIHA), the American College of Occupational and Environmental Medicine (ACOEM), the American Society of Safety Engineers (ASSE), and the American Association of Occupational Health Nurses (AAOHN). The committee also heard from the National Safety Council and representatives of business (Organization Resource Counselors, Inc.), labor (New York State Public Employees Federation, American Federation of Labor-Congress of Industrial Organizations) and the insurance industry (Liberty Mutual). Follow-up with the speakers provided more detailed information and points of contact for additional questions. The sponsors' project officers shared information on education and training from their files or put committee members in touch with the offices that had relevant data, and the committee members themselves contributed both personal contacts and specific information from their own files and experience. The World Wide Web provided much information about additional organizations and OSH training, and the following databases were accessed and searched: the National Center for Education Statistics database, National Center for Health Statistics Data Warehouse, the Federal Research in Progress database, the Federal Conference Papers database, Medline, MedStar, and HSRProj. The committee relied heavily on published and Internet-accessible data from the Bureau of Labor Statistics for projections of changes in the workforce and trends in occupational injuries and illnesses. The committee's discussion of distance learning (Chapter 8) was greatly enhanced by a commissioned paper on that topic written for the committee by Tim Stephens, the director of the Center for

Distance Learning at the University of North Carolina at Chapel Hill. This paper, and other written materials presented to the committee, are maintained by the Public Access Office of the National Research Council Library. Appointments to view these materials may be made by telephoning the library at (202) 334-3543 or by electronic mail to nrclib@nas.edu.

Because time, expense, and contract specifications ruled out collection of original survey data on both the supply of and the demand for OSH professionals, the committee drew on membership data from the leading OSH professional societies for its analysis of the current OSH workforce. AAOHN and the American Board of Occupational Health Nursing, ACOEM, AIHA, ASSE, and the Employee Assistance Professional Association all provided copies of membership demographics and recent member surveys on relevant topics. Other quantitative information describing the current OSH workforce came from the Internet sites of the Board of Certified Safety Professionals, the American Conference of Government Industrial Hygienists, the Human Factors and Ergonomics Society, and the Society for Industrial-Organizational Psychology.

ORGANIZATION OF THIS REPORT

Following this introductory chapter, Chapter 2 describes the current OSH professional workforce. Four chapters then describe current and anticipated trends in the general workforce, the workplace, the organization of work, and the delivery of health care. Chapter 3 explores the changing demographics of the United States and its possible effects on health and safety in the workplace. Chapter 4 examines the changing nature of work in the United States in the information age and the implications of those changes for occupational injuries and illnesses. Chapter 5 takes up the globalization of work in the 1990s and its consequences for workers and OSH professionals. In Chapter 6 the committee reviews a decade of changes in health care delivery in the United States, offers some thoughts on possible trends for the next decade, and points out the implications of those trends for the training of OSH personnel. Chapter 7 describes the current education and training programs that produce these OSH professionals. Chapter 8 analyzes the growing popularity of alternatives to campus-based, classroom instruction (e.g., "distance learning") and their application to the education and training of OSH personnel. The final chapter presents a brief summary and the committee's overall conclusions and recommendations.

2

Occupational Safety and Health Professionals

ABSTRACT. Without a massive survey of U.S. employers, it is impossible to estimate or describe the full spectrum of those who provide occupational safety and health (OSH) services to the U.S. workforce. However, it was possible to assemble a description of the four traditional or core OSH professions (occupational safety, industrial hygiene, occupational medicine, and occupational health nursing) as well as three other disciplines likely to play a substantial role in the workplace of the future: employee assistance professionals, ergonomists, and occupational health psychologists.

Although each of the four traditional OSH professions emphasizes different aspects of OSH, members of all four professions share the common goal of identifying hazardous conditions, materials, and practices in the workplace and assisting employers and workers in eliminating or reducing the attendant risks. Occupational safety professionals, although concerned about all workplace hazards, have traditionally emphasized the prevention of traumatic injuries and workplace fatalities. Similarly, industrial hygienists, although they do not ignore injuries, have been a source of special expertise on the identification and control of hazards associated with acute or chronic exposure to chemical, biological, and physical agents. Occupational health nurses and occupational medicine physicians are distinguished by providing clinical care and programs aimed at health promotion and protection and disease prevention. These services include not only diagnosis and treatment of work related illness and injury, but also pre-placement, periodic, and return-to-work examinations, impairment evaluations, independent medical examinations, drug testing, disability and case management, counseling for behavioral and emotional problems that affect job performance, and health screening and surveillance programs.

Approximately 76,000 Americans are active members of professional societies that represent the core OSH disciplines. The literature suggests that as many as 50,000 more are eligible for membership by virtue of their current employment. The committee therefore estimates the current supply of OSH professionals at 75,000 to 125,000. The committee could not locate good, independent data to support an estimate of demand (i.e., the number of available positions), but overall supply seems to be roughly consonant with employer demand. However, the committee notes that considerable need exists beyond the current demand for OSH professionals by employers. Doctoral-level safety educators are needed to maintain the supply of practicing safety professionals, and both occupational medicine and occupational health nursing need more specialists with formal training. Most importantly, a large fraction of the U.S. workforce is outside the sphere of influence of OSH professionals, particularly those whose focus is primarily prevention, principally because few are employed by small firms and establishments, and, in some sectors of the economy such as agriculture and construction, both the workplace and the workforce are transient.

Those who provide OSH services to the U.S. workforce are an extraordinarily diverse group (see Box 2-1). Every business has some safety or health hazards and should logically have someone responsible for the safety and health of its workers. Those vested with some degree of OSH responsibility range from medical specialists with residency training, who bring 22 years of education to bear on the task, to the workers themselves, who may have only a high school diploma and a few words of caution upon starting the job. Many individuals with significant responsibility have no formal training at all. Some come from fields like engineering, psychology, business, or one of the sciences and have highly relevant technical or professional education. Many others developed the relevant skills on the job, as full-time OSH specialists or as human resource managers or line supervisors with an additional duty as health and safety officer. Allied professionals include highly trained individuals who provide important health or safety services to the general population, which of course includes numerous workers (occupational therapists, audiologists, and orthopedic surgeons for example) or who provide such services in highly circumscribed settings (for example, infection control practitioners in hospitals or health physicists in industries where radiation is a hazard). Short of an exhaustive survey of U.S. businesses, it is impossible to estimate or describe the full spectrum of OSH personnel, but it is possible to construct a snapshot of those with formal education

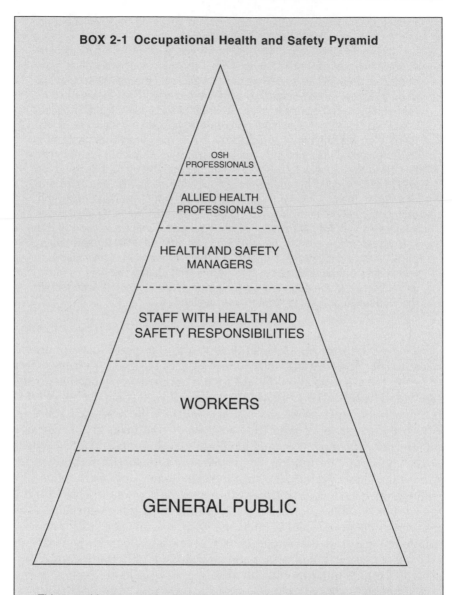

BOX 2-1 Occupational Health and Safety Pyramid

OSH
PROFESSIONALS

ALLIED HEALTH
PROFESSIONALS

HEALTH AND SAFETY
MANAGERS

STAFF WITH HEALTH AND
SAFETY RESPONSIBILITIES

WORKERS

GENERAL PUBLIC

This pyramid represents the sizes of various sectors of the population with de facto responsibility for OSH. It is not strictly proportional to the numbers involved, but the shape is intended to convey that the recognized professionals in the field constitute only a small portion of those involved. The triangular shape is also intended to convey the extent of contact with the general workforce. OSH professionals are concentrated in large firms, which employ a minority of U.S. workers.

and training. The traditional or core OSH professions are generally recognized as *occupational safety, industrial hygiene, occupational medicine (OM)*, and *occupational health nursing*. In fact, the National Institute for Occupational Safety and Health (NIOSH) education and training programs have focused almost exclusively on these professions since their inception in the late 1970s. The existence and interest of very active professional societies associated with these disciplines enabled the committee to assemble comprehensive views of each of these OSH fields, which are summarized in this chapter. The chapter also includes a short section that describes three other groups of professionals likely to play a substantial role in the workplace of the future: employee assistance professionals, ergonomists, and occupational health psychologists.

THE NATURE AND SCOPE OF OSH SERVICES

Although each of the four traditional OSH professions emphasizes different aspects of OSH and educates and trains its members accordingly, in practice, members of all four professions share the common goal of identifying hazardous conditions, materials, and practices in the workplace and assisting employers and workers in eliminating or reducing the attendant risks. Hazard identification requires knowledge of relevant laws and regulations; the physiological capabilities of workers; the materials, equipment, and processes in use at the work site; and historical data on the work site, the industry or business as a whole, and the individual workers at the work site. It may require interviews, surveys, environmental sampling, and laboratory analysis or the assistance of other professionals. Hazard control may involve engineering or design changes, procedural or administrative changes, provision of personal protective equipment, or changes in worker behavior. It almost always involves education and training of both management and workers about the hazard and its control.

Occupational safety professionals, although they are concerned about all workplace hazards, have traditionally emphasized the prevention of traumatic injuries and workplace fatalities. Similarly, industrial hygienists, although they do not ignore injuries, have been a source of special expertise on the acute and chronic effects of chemical, biological, and physical health hazards. Occupational health nurses and OM physicians, although they are concerned with hazard identification and control, are distinguished by providing clinical and preventive health care. These cover prevention, diagnosis, treatment, and referral, including preplacement, periodic, and return-to-work physicals, drug testing, disability management, counseling for behavioral and emotional problems that affect job performance, and health promotion and disease prevention programs.

In the biggest and best OSH programs, the special expertise of each is used in seamless coordination.

OSH Practice Settings

The settings in which OSH professionals practice are varied, but most can be classified into five major categories:

- industry and industry-like settings, including those associated with the military and government agencies,
- consulting firms including the insurance industry and some specialized government units,
- government regulatory agencies,
- educational and research institutions, and
- hospitals and outpatient clinics (nurses and physicians).

The traditional setting for much of OSH practice is in medium-sized or large industries where the OSH professional serves in a line function and addresses occupational health issues for a well-defined set of workers. The role is rather similar whether the industry is petroleum refining, a large bank, or an aircraft carrier. In all cases, the OSH professional focuses on the particular hazards of the industry and methods of their evaluation, control, and management. In some cases the OSH professional is assigned to a particular facility, and in others he or she operates from a corporate center.

An increasing number of OSH professionals work for consulting firms that provide OSH services to various segments of industry and government on a contractual basis. This includes those who work for insurance carriers that provide consulting services to the company's various clients. In some cases these relationships are stable and allow the development of industry-specific expertise, and in other cases the OSH practice is very broadly based and varied. Consulting practice presents considerable challenges in influencing internal corporate culture and mounting stable prevention activities from outside the company. Nevertheless, many companies are outsourcing OSH functions, particularly OM and industrial hygiene functions. The current mode of practice for OM physicians, for example, has changed from one that is dominated by physicians who are hired by large corporations to one in which the majority of OM physicians practice in the private sector as clinicians. As the size of the U.S. manufacturing sector has been reduced, the number of local work site-based physicians has also decreased. Instead, increasing numbers of OM physicians have established practices in hospital- and clinic-based health maintenance organizations and group practices or in private solo practice. Several companies have purchased individual clinics and have formed extensive networks of clinics that specialize in OM. An additional base from which OSH consulting activities are mounted are specialized government units, often linked to regulatory agencies, that assist small employers in addressing occupational and environmental health hazards.

The principal federal regulatory agency that makes extensive use of OSH expertise is the Occupational Safety and Health Administration (OSHA) of the U.S. Department of Labor. In addition, various states enforce the Occupational Safety and Health Administration Act of 1970 (OSHAct) under agreements with OSHA, and various local jurisdictions undertake regulatory activities. Other industry-specific regulatory programs that call on OSH skills and training, notably, in the Mine Safety and Health Administration (MSHA) of the U.S. Department of Labor and the Nuclear Regulatory Agency. The U.S. Department of Defense and the U.S. Department of Energy, which are self-regulating, also employ substantial numbers of OSH professionals, both in-house and as consultants and contractors. The goal of workplace inspections in all of these settings is to verify that the employer has accurately assessed and effectively controlled the hazards faced by its employees and to ensure that the workplace is in compliance with the appropriate regulations. The educational efforts of compliance officers are limited to informal on-site discussions to help employers understand the hazards and regulations that affect their workplaces and to inform workers and union representatives of employees' rights under the law.

Research in or directly relevant to the OSH field is carried out in government laboratories, notably, those of NIOSH, universities, and the private sector and in institutions affiliated with organized labor. Researchers are often affiliated with departments or units with "occupational safety and health" in their name, but many are not, which makes it difficult to estimate the size and extent of the research enterprise in the field. For example, aerosol science research, which is of great relevance to industrial hygiene practice, is often found in engineering departments in universities, as are many aspects of the control technology that underlie the control of workplace hazards. In contrast to the professional practice of occupational safety, industrial hygiene, OM, and occupational health nursing, many researchers who make important contributions to the field have had no OSH training. Scientists investigating the pathophysiology of cancer, or of asthma, for example, may know little about OSH but nevertheless provide information highly relevant for occupational risk assessments. This separation carries over into graduate degree programs in the field, as will be discussed below, in which the curriculum has both a highly structured technical component and a component that relates to the professional aspects of the field, including, for example, the federal regulatory system or the workers' compensation insurance system. University faculty members who contribute to OSH teaching programs also reflect this separation. Some have training and experience in the field, and others have technical knowledge of great relevance to the field but little professional OSH experience. The latter individuals often come from backgrounds in chemistry or engineering or they are physicians or nurses specializing in fields such as epidemiology, toxicology, respiratory disease, and dermatology. It is very important to foster the application of a wide variety of specialties and fields of knowledge to OSH problems. NIOSH's National Occupational Research Agenda illustrates the breadth and depth of the interdisciplinary research needs that must be addressed (National Institute for Occupational Safety and Health, 1998).

In addition to teaching at the college or university level, a number of OSH professionals hold other positions with a significant teaching component, either in continuing education for working professionals or in programs aimed at orienting management personnel or workers themselves to the field. Maintenance of certification in any of the four traditional OSH professions requires participation in continuing education courses, which provides incentives for developing courses and recruiting qualified instructors to meet this demand.

Environmental Health and Safety

It should be noted that although the focus of this report is on workplace health and safety, the knowledge and skills of OSH professionals are applicable outside the workplace as well. OSH professionals are equipped to deal with safety issues and physical, chemical, and biological hazards, wherever they occur, and the injuries and illnesses that they cause, whether in workers, consumers, or the general public. As a result, OSH personnel are increasingly required to address environmental health and safety issues. Manufacturers may ask their industrial hygienists to monitor not only the indoor air being breathed by employees but the level of hazardous emissions being released into the air and water of the surrounding community. Public health agencies or environmental groups may hire or otherwise call upon OSH professionals to monitor pollutants in community air and water as well. Occupational health clinics are prepared to diagnose and treat lead poisoning whether its patients are refinery workers or children eating lead-based paint in old houses. Environmental issues have become sufficiently important for physicians that the American College of Occupational Medicine changed its name to the American College of Occupational and Environmental Medicine and the specialty area is sometimes referred to as occupational medicine and sometimes as occupational/environmental medicine. The education and training of each of the four core OSH professions includes instruction on environmental health (see Chapter 7), and the occupational focus of this report should not be taken as ignorance of or a lack of appreciation for the importance of the contributions of OSH professionals to environmental health research and practice.

SAFETY PROFESSIONALS

The prevention of injuries, illness, and unexpected death for workers is the basic definition of occupational safety. Although occupational safety has historically focused on the prevention of acute traumatic injury, a broader definition generally includes the control of hazards and the prevention of accidents not only to protect the U.S. workforce but also to protect the general public and the environment. Therefore, the broad discipline of safety deals with the interaction between people and the physical, chemical, biological, and psychological effects, acute or chronic, that can adversely affect their well-being. The discipline of safety is the systematic application of principles drawn from engineering, physics, education, psychology, health and hygiene, enforcement, and management to prevent harm to people, property, and the environment.

The safety professional (SP) normally deals with the physical aspects of the workplace and their interaction with the worker and is directly concerned with injuries caused by slips and falls or by being struck by or crushed under an object, cuts, crushes, burns, electric shock, or improper lifting, bending, or stretching. The SP must be knowledgeable about the effects of all types of uncontrolled energy, such as electricity, pressures, weights, fluids, temperatures, motion or moving parts, radiation (ionizing and nonionizing), fires, and explosions. The SP must see that workers are issued and wear well-maintained personal protective equipment such as hard hats and helmets, goggles, safety shoes, respirators, clothing that protects individuals from hazardous chemicals, and the like.

The SP must understand and apply the OSHA General Industry, Construction, and Maritime standards, the American National Standards Institute and American Society for Testing and Materials standards, and, occasionally, international standards, as well as the specific standards of the mining, agriculture, and transportation sectors and those of product safety.

Safety Professional History

From the use and control of fire to early hunters protecting themselves from the hazards of wild beasts and reptiles, humans have recognized the need for occupational safety. Through the centuries, humans have recognized the myriad hazards that arise out of an increasingly industrializing world economy, such as the unexpected dangers of using power to fuel rapidly developing industries (see Appendix C for a more detailed list of the key events and individuals in the last two centuries). In the United States, before the mid-18th century, a high percentage of work was done on family farms, which consequently became the location for most worker injuries and fatalities. Later, with the onset of the Industrial Revolution in the 1800s, factories began to replace smaller shops and the changing work environment created a challenge for the prevention of job-related injuries, illness, and death.

In response to growing worker resentment toward the hazardous work conditions in factories, Massachusetts began using factory inspectors in 1867. Ten years later, additional legislation from Boston required the safeguarding of dangerous machinery. In the early 1900s, New Jersey, Wisconsin, and a number of other states enacted workers' compensation laws that made the employer financially liable for workplace accidents. With this incentive, organized safety programs were initiated and the SP came into existence. Initially, the SP was a person who assumed the responsibilities of carrying out the goals and objectives of a safety program. Only the larger and more progressive sectors, in particular, the steel and insurance industries, had a dedicated SP on staff. Other businesses assigned the task of preventing injuries to an experienced employee who knew the plant layout, equipment, and functions. These early SPs were also known as safety practitioners.

Safety Professional Services

The American Society of Safety Engineers (ASSE) identifies four primary functions of a safety professional (American Society of Safety Engineers, 1996):

1. Anticipate, identify, and evaluate hazardous conditions and practices. This function involves such activities as
- developing and applying methods for using experience, historical data, and other information sources to identify and predict hazards in existing or future systems, equipment, products, software, facilities, processes, operations, and procedures during their expected lifetimes;
- evaluating and assessing the probability and severity of losses and accidents that may result from actual or potential hazards;
- compiling and analyzing data from accident and loss reports and other sources to identify causes, trends, and relationships, ensure the completeness, accuracy, and validity of required in-

formation, evaluate the effectiveness of classification schemes and data collection methods, and initiate investigations;

- providing advice and counsel about compliance with safety, health, and environmental laws, codes, regulations, and standards;
- conducting research studies of existing or potential safety and health problems and issues; and
- conducting surveys and appraisals to identify conditions or practices that require the services of specialists such as physicians, health physicists, industrial hygienists, fire protection engineers, design and process engineers, ergonomists, risk managers, environmental professionals, psychologists, and others.

2. Develop hazard control designs, methods, procedures, and programs. This function involves such activities as

- formulating and prescribing engineering or administrative controls to eliminate hazards, exposures, accidents, and losses and to reduce the probability or severity of injuries, illnesses, and losses when hazards cannot be eliminated;
- devising methods to integrate safety performance into the goals and operations of organizations and their management systems; and
- developing safety, health, and environmental policies, procedures, codes, and standards for integration into operational policies of organizations, unit operations, purchasing, and contracting.

3. Implement, administer, and advise others on hazard controls and hazard control programs. This often entails

- preparing valid and comprehensive recommendations for hazard controls and hazard control policies, procedures, and programs that are based on analysis and interpretation of accident, exposure, loss event, and other data;
- directing or assisting in developing educational and training materials or courses;
- conducting or assisting with courses related to hazard recognition and control;
- advising others about communicating with the media, community, and public about hazards, hazard controls, relative risk, and related safety matters; and
- managing and implementing hazard controls and hazard control programs.

4. Measure, audit, and evaluate the effectiveness of hazard controls and hazard control programs. This function involves

- establishing techniques for risk analysis, cost-benefit analysis, work sampling, loss rate, and similar methodologies;
- developing methods to evaluate the costs and effectiveness of hazard controls and programs; and
- directing, developing, or helping to develop management accountability and audit programs.

Specific roles and responsibilities vary widely, depending on the education and experience of the individual and the nature of the organization that employs him or her. SPs with a doctoral degree can be found teaching and doing research at colleges and universities, performing public service, and consulting. Most SPs have bachelor's or master's degrees, however, and work for insurance companies, a wide variety of industries, state and federal agencies such as OSHA, hospitals, schools, and non-profit organizations.

Safety Professional Education

A number of community and junior colleges offer 2-year programs that lead to an associate degree or a certificate in safety or a related field. Some of these programs are designed to prepare students to enter the workforce as safety technologists, and others prepare students for transfer to a 4-year safety degree program. Over the last decade about 50 percent of those in the safety field have held a bachelor's degree as their highest degree. About 30 percent of those who enter the field have a bachelor's degree in safety, but many move into safety from other disciplines (e.g., engineering, business, and physical sciences) and later pursue safety studies.

More than 30 colleges and universities offer a bachelor of science in safety. However, only six institutions, none of which receive support from NIOSH, offer safety degree programs accredited by the Accreditation Board for Engineering and Technology (ABET). Requirements for a major in safety typically include courses on safety and health program management, design of engineering hazard controls, system safety, industrial hygiene and toxicology, accident investigation, product safety, construction safety, fire protection, ergonomics, educational and training methods, and behavioral aspects of safety. To prepare for these courses, students are generally required to take courses in mathematics, chemistry, physics, biology, statistics, business, engineering, and psychology. Good computer skills are a necessity as well. Many safety degree programs provide opportunities for students to work with SPs at local companies as interns.

A 1997 survey of ASSE members found that 28 percent reported a master's degree as their highest degree and 3 percent reported a doctoral degree as their highest degree (American Society of Safety Engineers, 1997). Only 17 percent reported less than a bachelor's degree. Twenty-nine universities offer master's degree graduate programs in safety, most of which may involve specialization in fields such as management, engineering, ergonomics, loss control, risk management, and fire protection or in industries like agriculture, mining, construction, and chemical processes. Only five of these programs are currently accredited by ABET. Doctoral programs in safety are offered by nine universities (ABET does not accredit doctoral programs), but analysis of limited data provided to the committee by NIOSH grantees showed that in the previous 5 years only one dissertation focused on acute injury prevention generally instead of one of the subspecialties listed above.

Safety Professional Certification

To date no state has required safety professionals to be licensed, but the safety profession has established its own program to provide actual and potential employers some means of assessing professional competency. The Board of Certified Safety Professionals (BCSP) was established in 1969 to evaluate candidates and offer the Certified Safety Professional (CSP) and Associate Safety Professional (ASP) designations to those who meet its standards. Under the sponsorship of ASSE, the American Industrial Hygiene Association (AIHA), the System Safety Society, the Society of Fire Protection Engineers, the National Safety Council, and the Institute of Industrial Engineers, BCSP evaluates the academic and professional experience qualifications of safety professionals, administers examinations, and issues certificates of qualification to those professionals who meet the Board's criteria and successfully pass its examinations.

The preferred qualifications for certification are a bachelor's degree in safety from a program accredited by ABET, 4 years of professional safety practice, and passing of two examinations given by BCSP. Because many people enter the safety profession from other educational backgrounds, candidates for certification may substitute other degrees plus professional safety experience for an accredited bachelor's degree in safety. BCSP uses a point system to determine examination eligibility, but minimum educational qualifications are an associate degree in safety and health or a bachelor's degree in any field. Continuing education courses, seminars, and certificate programs do not receive credit for the academic requirement, but each month of acceptable professional safety experience earns one point. The total points are the sum of academic points and experience points.

In addition to the academic requirement, CSP candidates must have 4 years of professional safety experience. The 4 years are in addition to any experience used to meet the academic requirement. Professional safety experience must meet all of the following criteria to be considered acceptable:

- The professional safety function must be the primary function (50 percent) of the position.
- The position's primary responsibility must be the prevention of harm rather than responsibility for responding to harmful events.
- The position must be full time (at least 35 hours per week).
- The position must be at the professional level.
- The position must have a breadth of duties.

The process used to achieve the CSP designation typically involves the passing of two examinations: one on safety fundamentals and a second on comprehensive practice. The Safety Fundamentals Examination, which covers basic knowledge, is taken first. Upon passing of that examination, candidates receive the ASP title to denote their progress toward the CSP. Some candidates who have been examined through other acceptable certification and licensing programs and who currently hold such certifications or licenses may waive the Safety Fundamentals Examination. The acceptable certifications or licenses most relevant to this report are the Certified Industrial Hygienist (CIH) from the American Board of Industrial Hygiene, the Registered Professional Engineer (RPE) from the engineering registration board for any U.S. state or territory, and the Certified Hazard Control Manager (CHCM) from the Institute of Hazardous Material Management.

To take the Comprehensive Practice Examination, a candidate must meet more demanding academic and experience requirements and must have passed or been waived from the Safety Fundamentals Examination. After passing the Comprehensive Practice Examination, a candidate receives the CSP title. At 5-year intervals thereafter, the individual is recertified, contingent upon accumulation of 25 "continuation of certification points" through approved continuing education courses. ASSE salary surveys indicate that individuals with the CSP designation typically earn 15 to 20 percent more than their uncertified counterparts. As of 1998, approximately 10,000 individuals held the CSP designation, and about 2,000 held the ASP designation.

Current Status of the Safety Professional Workforce

As is the case with the other core OSH disciplines, the committee focused on membership in the leading professional societies as a method of estimating the size and composition of the safety workforce, although it recognized that this surrogate measure is unlikely to capture all those currently functioning as safety practitioners and may well lead to double counting of a substantial number of OSH professionals. This section presents the basic data on SP society members, followed by similar data on those SPs who have earned the CSP designation.

In the case of SPs, the leading professional society is ASSE, which currently has a membership of approximately 33,000. Membership requires current employment in safety or one of its relevant specialties. About 10,000 members are designated "professional members" by virtue of certification and 5 years of safety experience or by virtue of a bachelor's degree from an accredited college or university and 10 years of safety experience. A 1997 survey of the membership (American Society of Safety Engineers, 1997) revealed that the professional members who responded had worked in the safety field for an average of 19 years. Seventy-one percent had worked in the field for more than 15 years.

Almost 80 percent of the professional members of ASSE are certified: 66 percent hold the CSP designation, and another 13 percent are certified by another safety-related or engineering organization. Comparison with previous surveys in 1981 and 1990 shows that the proportion of profes-

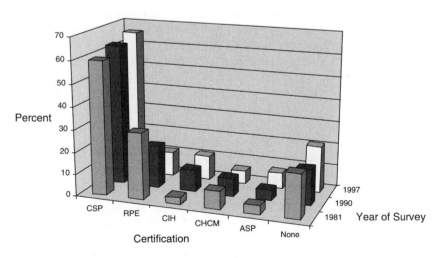

FIGURE 2-1 Percentage of professional members reporting safety-related certifications in three surveys. SOURCE: American Society of Safety Engineers (1997).

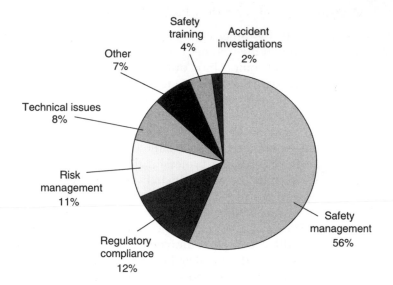

FIGURE 2–2 Primary job focus of ASSE members. SOURCE: American Society of Safety Engineers (1997).

sional members with the CSP designation has slowly increased and that the proportion of members who report themselves as Registered Professional Engineers decreased from 30 percent in 1981 to 19 percent in 1990 to 11 percent in 1997 (see Figure 2-1).

The ASSE survey also asked members how they actually spend their time at work. Not surprisingly, three-fourths of the respondents indicated that they spent at least 50 percent of their time on safety and health-oriented issues. On average, respondents reported that they spent 59 percent of their time on safety, 10 percent on health issues, 10 percent on industrial hygiene, 9 percent on environmental issues, and 12 percent on other activities. Figure 2-2 shows the answers when respondents were asked to choose their "primary job focus" from more a specific listing.

According to this 1997 ASSE survey, an SP spends more than half of his or her time (56 percent) carrying out responsibilities identified as safety management. This includes establishing the safety policy for the enterprise, large or small. The SP generally has the responsibility for establishing a hazard control system with goals, objectives, evaluations, and feedback. In larger businesses a staff of safety practitioners could be supervised. The SP can be an ex-officio member of various safety and health committees that focus on prevention activities. Other activities under the rubric of safety management are safety office finances, public awareness, off-the-job safety, recognition and rewards for safe practices, workplace task analyses, direction of accident investigations, hiring of safety and health consultants, documentation of near misses, injuries, illnesses, and fatalities, and application of intervention techniques.

The SP devotes a considerable amount of time, approximately 12 percent, ensuring that the place of work is in compliance with all of the applicable OSHA (federal and state), MSHA, and U.S. Department of Transportation standards. Another significant percentage of the SP's time (11 percent) is spent in the realm of risk management. This is best defined as the professional assessment of all potential sources of loss in an organization's structure and operations, and the selection of actions that will reduce losses to an acceptable minimum at the lowest possible cost. Such actions encompass the selection of insurance and the limits of self-insurance and assumed risk.

The survey respondents report that SPs devote about 8 percent of their time to dealing with technical issues. These are site specific, but they include such actions as proper guarding of machinery and equipment, safe storage of flammable and explosive materials, implementation of proper lifting techniques, and coping with workplace violence.

Even though the training of management and employees is of the utmost importance, the survey data show that SPs spend less than 5 percent of their time carrying out this responsibility, which involves determining what training is needed, setting the goals and objectives of the training, determining what it will take to carry out the training, conducting or seeing to it that the training takes place, evaluating the effectiveness of the training, and gathering the necessary feedback to improve future training. Ideally, each supervisor would be responsible for the instruction of the workers under his or her direction, but the SP must usually take the initiative to ensure that work is performed without undue risk and that it complies with more than 200 OSHA standards that specifically require employees to be trained in some aspect of occupational safety and health.

Figure 2-2 also indicates that a SP devotes 2 percent of her or his time to investigating the facts and background information surrounding workplace accidents to determine what caused the incident and how to prevent future occurrences. The range of severity of accidents can be from a slight cut or bruise to one in which there are multiple fatalities and a situation of multiple and complex causes, such as a crane collapse or a grain elevator explosion.

An important activity presumably subsumed by the very general categories of the survey is keeping required records. These would include logs for OSHA and the Bureau of Labor Statistics; workers' compensation records; records on costs, training, equipment inspection, personal pro-

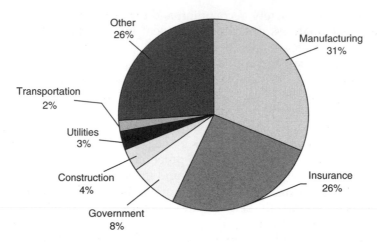

FIGURE 2-3 Employers reported by certified safety professionals (n = 9,848) in a 1997 survey. SOURCE: American Society of Safety Engineers and Board of Certified Safety Professionals (1997).

tective equipment, exposure to chemicals and radiation, and hazardous materials inventory; and on occasion, medical records. Others activities not explicitly listed above are providing first aid, holding fire and evacuation drills, and providing vehicular safety and legislative awareness activities. Frequently, an SP in a small or midsize business must assume responsibilities in such areas as industrial hygiene, ergonomics, radiological health, hazardous waste, and the environment.

A survey of more than 9,000 CSPs by BCSP provides detailed information about employers (American Society of Safety Engineers and Board of Certified Safety Professionals, 1997). Figure 2-3 shows that 31 percent of these respondents indicated that they worked in manufacturing or production industries. "Chemicals and allied products" accounted for a quarter of these, and "crude petroleum and natural gas" and "petroleum refining and related industries" together accounted for an additional 20 percent. None of the other 20 industries was named by more than 10 percent of respondents.

Of the 8.2 percent of CSPs employed by government, nearly two-thirds (65 percent) were employed by the federal government. State governments accounted for 21 percent, and local governments accounted for 13 percent. Consulting was by far the most common response in the large (26 percent) "Other" category, accounting for 53.7 percent of the respondents in that group. Education (colleges, universities, libraries) was a distant second with 12.2 percent.

INDUSTRIAL HYGIENE

Industrial hygiene is a field of applied science that concerns the anticipation, recognition, evaluation, and control of environmental factors in the workplace whose effects on workers can range from discomfort to chronic disease. Although the recognition and evaluation functions draw significantly from the biological sciences, other aspects of evaluation and the control function draw principally from the physical sciences and engineering. As a consequence, most industrial hygienists come from an educational background in the natural sciences or engineering and have substantial postgraduate training, either in formal degree programs or in continuing education courses.

Industrial Hygiene History

Appendix C provides a list of events and people who were important in the development of the OSH field as a whole, so this section focuses on some events and people of particular significance for industrial hygiene. The establishment of the Division of Industrial Hygiene in the Labor Department of the State of New York and the Office of Industrial Hygiene and Sanitation of the U.S. Public Health Service in 1913 (Corn and Corn, 1988) and the Industrial Hygiene Section of the American Public Health Association in 1914 signaled the beginnings of industrial hygiene as a distinct professional field (Smyth, 1966). By 1917, research on methods of measuring hazardous concentrations of carbon monoxide in air was under way at Harvard University (Corn and Corn, 1988), and the academic connection of industrial hygiene in both training and research has been maintained since that time.

Activities in industrial hygiene expanded throughout the 1920s and 1930s, a period that saw the development of professional guidelines for limiting occupational exposures to hazardous airborne agents. At that time they were called maximum allowable concentrations (MACs). In 1938 the American Conference of Governmental Industrial Hygienists (ACGIH) was founded with a membership of 76. ACGIH assumed responsibility for MACs during the 1940s and has had a major worldwide impact through the MACs and their successor guidelines, the threshold limit values (Corn and Corn, 1988). In 1939 AIHA was established to serve the professional interests of the estimated 300 industrial hygienists active in the United States at that time (American Industrial Hygiene Association, 1994). Today AIHA has 68 local sections and a membership of approximately 13,000, including 1,000 foreign members (Donald Ethier, AIHA, personal communication, August 9, 1999). ACGIH, with its more restricted membership, had 5,012 members in 1998, of which about 800 were foreign members (American Conference of Governmental Industrial Hygienists, 1999).

By the mid-1950s there was a recognized need for formalization of the professional practice of industrial hygiene, which culminated in the formation of the American Board of Industrial Hygiene (ABIH) in 1960 (Smyth, 1966). The purpose of the board was to develop and administer certification procedures that recognize "special education in and knowledge of the basic principles of industrial hygiene" (American Board of Industrial Hygiene, 1998).

Industrial Hygienist Services

The typical activities carried out by industrial hygienists are summarized by the ABIH (1998) as follows:

- review projects, designs, and purchases to anticipate health hazards;
- critically evaluate work environments, processes, materials inventories, and worker demographics to recognize potential health risks to persons or communities;
- assess human exposures to hazards through a combination of qualitative and quantitative methods to determine health risks, regulatory compliance, and legal liabilities;
- recommend effective control measures to mitigate risks via engineering, administrative, or personal protective methods;
- communicate risks and control measures to workers, management, clients, customers, and communities;
- provide specific training to workers about risks and control measures;
- perform laboratory analysis of samples taken to assess worker exposure;
- conduct research and development on industrial hygiene methods and tools;
- interface industrial hygiene programs with related health risk management efforts, including safety, environmental protection, and medicine;
- interface with regulatory, community, and professional organizations;
- manage, supervise, or advise other industrial hygiene staff;
- manage and advocate industrial hygiene programs;
- audit industrial hygiene programs;
- provide technical support to legal proceedings in matters related to industrial hygiene; and

• provide academic training in industrial hygiene at the college or university level.

Industrial Hygienist Education

A master's degree is the most common entry-level degree for those entering the field of industrial hygiene without prior work experience. Generally, a bachelor of science degree in one of the natural sciences or engineering is the preferred curricular prerequisite. For the most part, a master's degree with a specialization in industrial hygiene is offered in schools of public health. Those programs based in schools of public health sometimes offer curricula that lead to a Master of Public Health degree or a Master of Science degree, with the former demanding broader curricular content in public health. Common elements of all of these curricula for a master's degree are course work in the evaluation and control of chemical, biological, and physical hazards in the workplace, control technology including industrial ventilation, and ergonomics and an introduction to the professional and regulatory aspects of industrial hygiene practice. These are generally 2-year programs, with the intervening summer devoted to either a research project or an internship in an industrial or governmental setting.

In the early 1990s, an accreditation program was established for both bachelor's and master's degree programs in industrial hygiene under the auspices of ABET. Because graduation from an ABET-accredited program presents some advantage for individuals who intend to pursue industrial hygiene certification, this has been an inducement for college and university degree programs to become accredited. In 1999 there were 5 ABET-accredited bachelor's degree programs and 26 ABET-accredited master's degree programs (Accreditation Board for Engineering and Technology, 1999). An additional 30 to 35 master's degree programs have not yet been accredited by ABET. Collectively, these programs award approximately 400 master's degrees annually.

Most of the research universities that offer master's degree programs in industrial hygiene also offer doctoral degree programs with further training and research opportunities in the field. However, dissertation research directly related to industrial hygiene is also carried out, for example, in engineering departments, chemistry departments, and public health programs that are not identified in their name as supporting advanced education in the field of industrial hygiene. As a result, it is common for the industrial hygiene faculty at the research universities to be a mixture of those trained in industrial hygiene explicitly and others with doctoral training in engineering, chemistry, or other natural sciences.

NIOSH has been actively supporting postgraduate programs in industrial hygiene since its inception. In 1999 NIOSH provided training grant assistance to 29 industrial hygiene programs throughout the United States, 15 of which were at NIOSH-sponsored Education and Research Centers (ERCs). All but 1 of the 29 programs are at the postgraduate level, and collectively they produce approximately 200 master's-level industrial hygienists annually.

Continuing Education in Industrial Hygiene

Continuing education courses are a major feature of professional training in industrial hygiene. This is due in part to the requirement for annual participation in continuing education to maintain ABIH certification in industrial hygiene. Continuing education courses also provide a means of entry into the field for those unable to devote several years to graduate education as well as provide an orientation to the field for those with closely related responsibilities in medium-sized to large organizations. As will be reported in more detail in Chapter 7, continuing education is a second major component of NIOSH training programs. In addition to the NIOSH-supported institutions, continuing education courses in industrial hygiene are provided by the extension branches of various universities under the auspices of AIHA and by private firms.

Certification of Industrial Hygienists

As noted above, ABIH was founded in 1960 to develop and administer certification procedures. Certification involves two written examinations. The first is a 1-day core examination in basic industrial hygiene practice. The second is either the comprehensive practice or chemical practice examination, depending on the candidate's qualifications.

The core examination covers areas of industrial hygiene in which all industrial hygienists should be knowledgeable. All candidates for certification must take and pass this examination. Successful completion of the core examination entitles the applicant to a certificate as an industrial hygienist in training (IHIT). Persons who have recognized skills and expertise in industrial hygiene chemistry and the required experience may be approved for certification in chemical practice. Persons engaged in the general practice of industrial hygiene usually choose to take the comprehensive practice examination, an examination that covers all aspects of practice in greater depth than the core examination. ABIH examinations are conducted in the spring of each year at the American Industrial Hygiene Conference and Exposition. Examinations are also held each October, generally in about 10 cities throughout the United States and Canada, and at the Professional Conference on Industrial Hygiene.

Certain prerequisites must be met to take each of these examinations. An acceptable baccalaureate degree in engineering or a science from a college or university acceptable to ABIH and 1 year of professional industrial hygiene practice are required for admission to the core examination. The baccalaureate degree and a minimum of 5 years of professional practice are required for admission to the examination in comprehensive practice or chemical practice of industrial hygiene. ABIH may allow up to a maximum of 2 years of total credit for a completed graduate degree in lieu of part of the required experience. The allowable credit is 1 year of experience for an acceptable master's degree or 2 years for an acceptable doctoral degree.

Upon successful completion of the core and chemical practice or comprehensive practice examinations, candidates become diplomates of ABIH and members of the American Academy of Industrial Hygiene, a professional society whose membership is exclusively composed of certified industrial hygienists. They also receive an official certificate and individual certification number. They are authorized to use the title Certified Industrial Hygienist and the designation CIH. They are then responsible for the continued maintenance of their certification.

Exactly half of the 502 individuals who took the core examination in 1998 passed, and a little over one-third of those sitting for the comprehensive practice test passed that exam. Only nine people took the chemical practice examination (four passed). As of May 1999, ABIH listed 6,356 CIHs in active status and 643 IHITs. Among the CIHs, 6,005 had passed the comprehensive practice examination and 284 had passed the chemical practice examination.

As in a number of other professions, industrial hygienists must show continued professional qualifications. Beginning in 1979, CIHs have been required to demonstrate their continued professional development on a 6-year cycle. Various activities have been accepted by ABIH as evidence of continued professional qualification for certification maintenance. These include continuing professional industrial hygiene practice; membership in approved professional societies; teaching; publication in peer-reviewed journals; participation on technical committees; attendance at approved meetings, seminars, and short courses; approved extracurricular professional activities; or reexamination. Points are determined and awarded by ABIH for each approved activity.

Current Status of the Industrial Hygiene Workforce

As is generally the case with a diverse and geographically dispersed workforce, it is difficult to acquire a comprehensive view of the demographics of the current industrial hygiene workforce, its deployment in the U.S. economy, or those changes that it is experiencing in response to the changes in the organization and structure of work itself. The principal source of information available on these issues relating to the field of industrial hygiene comes from AIHA, especially a sample survey of its members conducted in 1997 (see below). In August 1999 AIHA listed 8,800 full members, that is, graduates of an accredited college with a baccalaureate or graduate degree in industrial hygiene, chemistry, physics, engineering, biology, or related discipline who have been engaged a majority of the time for at least 3 years in industrial hygiene-related activities. Its 1,400 associate members are individuals who are otherwise qualified but who have less than 3 years of experience in the industrial hygiene field. Student, retiree, affiliate, and organizational members bring the total to a little more than 12,000.

The 1997 Definition of the Profession Survey (Association Research Inc., 1997) randomly selected 2,000 members who were sent comprehensive questionnaires concerning their education and professional practice. Keeping in mind the potential biases that might be introduced by unknown differences between the 915 members who responded and those who did not, as well as the representativeness of AIHA members among all industrial hygienists, the following paragraphs summarize some of the results of that survey.

The respondents to the survey were employed in private industry (47 percent), consulting firms (17 percent), government (11 percent), and academia (6 percent). Another 6 percent were self-employed, with the remainder falling into the "other" category. Almost 75 percent had been with their current employer for 5 years or more. The respondents were rather uniformly distributed across the categories of years in the profession, which were less than 15 years, 15 to 19 years, 20 to 24 years, and 25 years or more. Eighty percent of the respondents were male, and female respondents were generally much younger. It is interesting that 57 percent of the respondents did not begin their careers in the field of industrial hygiene, but younger respondents were more likely to have done so. Eighty-three percent of respondents had been members of AIHA for 10 years or more, and 80 percent were CIHs. The highest degrees held were bachelor's (25 percent), master's (61 percent), and doctoral (11 percent) degrees. Hence, the respondents are highly educated and experienced, long-term professionals.

The experience of the respondents was reflected by the fact that more than one-third stated that their primary responsibility was for program management or administration. Another 25 percent operated principally as consultants, and only 6 percent were in education or research. Only a small fraction of the respondents had primary responsibility for safety (6 percent) or environmental concerns (2 percent). The degree to which hygienists were responsible for safety and environmental concerns was, however, an issue of some interest, and members were asked various questions about this issue. The overall picture that emerges is that hygienists currently spend modest amounts of time on safety (19 percent) and environmental issues (12 percent) but that there is a widely shared expectation (82 percent) that these areas will become more integrated with industrial hygiene in institutional planning and management in the future. The implications for industrial hygiene practice are that there will be less technical specialization and more involvement in a broader range of issues, for example, international matters and employee education. A more detailed picture of what respondents do on the job is presented in Table 2-1, which indicates the activities most commonly chosen as one of the "top three most important activities worked on during the year." Responses to another question indicated that, on average, 43.7 percent of respondents' time was spent on the most important activity, 20.6 percent was spent on the second most important activity, and 14.7 percent was spent on the third most important activity.

Another view of the nature of their jobs is shown in the next two tables, which indicate the activities in which respondents have hands-on involvement (Table 2-2) and the activities most likely to be managed by respondents (Table 2-3).

For the most part, Tables 2-1 to 2-3 show that respondents are engaged in quite traditional industrial hygiene activities, with a few newer areas being added, for example, indoor air quality and risk communica-

TABLE 2-1 Percentage of Respondents Reporting the Indicated Activities as Among the Three Most Important Worked on During the Preceding Year

Activity	Percentage
Administration or management	37.8
Chemical sampling	35.1
Industrial hygiene program documentation	19.2
Managing safety programs	17.2
Risk assessment	16.3
Employee safety training	14.0
Indoor air quality assessment	14.0
Occupational health	13.5
Physical agent monitoring	13.3
Environmental activities	12.7

SOURCE: Association Research, Inc. (1997).

TABLE 2-2 Percentage of Respondents Reporting Hands-on Involvement in the Indicated Activities

Activity	Percentage
Indoor air quality evaluation	45.6
Risk communication	45.5
Personal protective equipment evaluation	45.5
Risk assessment	44.5
Hazard analysis	44.4
Chemical sampling	44.3
Physical agent monitoring	44.1
Industrial hygiene program evaluation	41.3

SOURCE: Association Research, Inc. (1997).

TABLE 2-3 Percentage of Respondents Reporting That They Manage the Indicated Activities

Activity	Percentage
Chemical sampling	53.7
Industrial hygiene program documentation	51.1
Asbestos	50.8
Physical agent monitoring	49.2
Personal protective equipment evaluation	46.9
Lead	46.5
Confined spaces	46.4
Occupational health	45.1

SOURCE: Association Research, Inc. (1997).

tion. Notable for its absence in all of these lists is the burgeoning area of ergonomics, which is now a common element of industrial hygiene curricula as well as of industrial hygiene practice.

A particularly interesting finding arose from questions that asked with whom the respondents interacted as a part of their daily routine. The responses are shown in Figure 2-4, which indicates a high level of interaction with the management structure as well as with workers. It is these interactions that are significantly diminished by outsourcing the industrial hygiene function in medium-sized to large corporations and why there is concern about this trend in the profession.

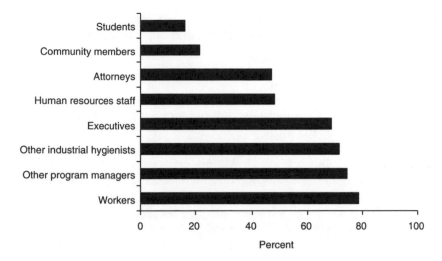

FIGURE 2-4 Percentage of respondents reporting daily direct interaction with the indicated groups. SOURCE: Association Research, Inc. (1997).

It is also interesting that almost half of the respondents interact routinely with attorneys but that only one in five interact with either community members or students. One explanation for this pattern of interactions is that the respondents were mostly relatively senior and experienced individuals and therefore tended to function principally in an administrative role. However, worker interaction was still the most common daily interaction, and substantial involvement in hands-on activities was reported.

OCCUPATIONAL MEDICINE

OM is the area of preventive medicine that focuses on the relationships among the health of workers, the ability to perform work, the arrangements of work, and the physical, chemical, and social environments of the workplace. Practitioners in this field recognize that work and the environment in which work is performed can have favorable or adverse effects on the health of workers as well as on that of other populations, that the nature or circumstances of work can be arranged to protect worker health, and that health and well-being in the workplace are promoted when workers' physical attributes or limitations are accommodated in job placement (Accreditation Council for Graduate Medical Education, 1999). OM specialists, who are often involved in direct patient care, identify and control work-related disease and injury and seek ways to eliminate and reduce hazards in factories, mines, offices, and other work settings.

Occupational Medicine History

The reader is again referred to Appendix C for a fuller list of important events in OSH. The following few paragraphs provide some additional context specific to OM.

Bernardino Ramazzini (1633–1714) is considered the father of OM in the western world. In his book *De Morbis Artificum Diatriba,* Ramazzini described diseases of the trades. Ramazzini instructed his students to ask one additional question to patients: "What is your occupation?" (Levenstein, 1988).

Alice Hamilton (1869–1970) is considered the matriarch of OM in the United States. She was the first U.S. physician to devote her professional life to the practice of OM (Bendiner, 1995). While living at Hull House in Chicago in 1908, Hamilton exposed an epidemic of "phossy jaw" caused by exposures to white phosphorus among matchmaking workers. Through her efforts, U.S. manufacturers changed to a safer type of organic phosphorus to make matches (Levenstein, 1988; Hamilton, 1990). The passage of workers' compensation acts by the states and the federal government from 1909 to 1911 changed how industry had to view work-related injuries and illnesses. This was the beginning of the acceptance by industry that it had to both control workplace hazards and provide medical care to injured workers (Rom, 1992).

The American Association of Industrial Physicians was founded in 1916. The organization was renamed the Industrial Medical Association and finally the American Occupational Medicine Association (AOMA). This was primarily the professional organization for physicians employed by industry. In 1946 the American Academy of Occupational Medicine (AAOM) was founded. AAOM was more of an academic organization than AOMA was. In 1988, AOMA and AAOM merged and, in acknowledgment of the increasing importance of environmental medicine, formed the American College of Occupational and Environmental Medicine (ACOEM) (Rom, 1992).

Various diseases caused by exposures to agents such as white phosphorus, silica, asbestos, vinyl chloride, and coal dust have caused the most public attention (many of the OM physicians during the first half of this century were also industrial hygienists or health physicists and played important roles in the establishment of those professions). However, the primary focus of OM practice in the latter half of the 20th century was trauma or industrial injuries, which have historically generated the lion's share of workers' compensation claims and payments.

Starting with the passage of the state-based workers' compensation acts, companies began hiring physicians to be plant physicians. One of the issues that confronted the American Association of Industrial Physicians was the role and scope of these activities. Community physicians were concerned that the industrial physicians might "steal" their patients. In some other countries health care has in fact been organized around the workplace—the same physicians take care of both the workers and their families for all conditions, both work-related and non-work-related conditions. This concern of U.S. community physicians led to an understanding regarding the role and scope of industrial physician activities: they would treat industry-caused illnesses and injuries but would not treat non-industry-caused conditions. This understanding has allowed a generally harmonious relationship between the industrial physicians and their community counterparts.

In 1955, the American Board of Preventive Medicine recognized OM as one of the three disciplines of preventive medicine. Since then some 2,200 physicians have been acknowledged as board certified in OM (American Board of Preventive Medicine, 1999), but relatively few physicians were formally trained in OM before the passage of the OSHAct of 1970. Because the training programs, like all preventive medicine programs, do not generate much revenue from patient treatment, money to support residents was (and still is) in short supply (Anstadt, 1999). Because the field was considered a subspecialty of preventive medicine, most of the 40 training programs are associated with universities with public health schools or specific departments within a medical school.

Occupational Medicine Services

ACOEM has developed an extensive list of "competencies" describing specific behaviors characteristic of good OM practice (Upfal et al., 1998). Upfal and colleagues recognize that practices vary widely, that few if any practitioners will be expert in all of the listed competencies, and that each practitioner will have a unique spectrum of competencies. All OM physicians will have a strong clinical emphasis and will be familiar with issues of worker placement and accommodation. In addition to this grounding in clinical practice, subspecialists will have competencies in public health, prevention, population medicine, epidemiology, toxicology, and research methods, as well as other competencies on the list. The list is too extensive to print here in its entirety, but the following introductory excerpts provide a general summary (Upfal et al., 1998):

- **Clinical—General.** Physicians with competency in this area have the clinical knowledge and skills required to provide high-quality, cost-effective medical care in diagnosing and treating occupational and environmental injuries and illnesses. The physician provides care with an understanding of the workplace, work exposures, and relevant statutes, such as workers' compensation laws. Throughout the course of care, the physician seeks to maximize the patient's functional recovery. The physician also seeks to identify and reduce workplace and environmental hazards to reduce the risk of future injury or illness to the patient.

- **Clinical—Preventive.** Physicians with competency in this area have the knowledge and skills required to define, develop, and administer programs to improve the health of employee and dependent populations, as well as counsel employees about their lifestyle risk factors and clinical preventive needs. The physician is able to apply a full range of primary, secondary, and tertiary preventive methods to this end.

- **Public Health and Surveillance.** Physicians with competency in this area have the knowledge and skills required to recognize and address conditions of public health importance, with an emphasis on prevention, as well as to monitor populations for indicators of occupational or environmental health effects.

- **Disability Management and Work Fitness.** Physicians with competency in this area have the clinical and administrative knowledge and skills required to assist employees and employers to ensure that recovery from illness or injury is as rapid and as complete as possible. With broad knowledge of the workplace, administrative requirements governing job placement, and the legal, rehabilitative, and financial aspects of disability, the occupational and environmental physician facilitates the restoration of productivity for the injured or ill employee and assesses safe work capacities to permit work placements that safeguard employees and others.

- **Hazard Recognition, Evaluation, and Control.** Physicians with competency in this area have the knowledge and skills required to (1) recognize and evaluate or assist in evaluating potentially hazardous workplace and environmental conditions, (2) recommend or implement controls or programs to reduce such exposures, and (3) evaluate the impacts of such exposures on the health of individual workers, patients, and the public. The physician collaborates with other professionals, such as industrial hygienists, safety engineers, ergonomists, and occupational health nurses, on such efforts.

- **Regulations and Government Agencies.** Physicians with competency in this area have the knowledge and skills required to help bring organizations into compliance with state and federal regulations relating to OEM as well as general public health laws.

- **Management and Administration.** Physicians with competency in this have the administrative and management knowledge and skills required to plan, design, implement, manage, and evaluate comprehensive occupational and environmental health programs, projects, and protocols that enhance the health, safety, and productivity of workers, their families, and members of the community. The spectrum of activities may vary substantially depending upon the physician's practice setting and the characteristics of the organization(s) served.

Occupational Medicine Education and Training

By definition, OM physicians must possess a medical degree (M.D. or D.O.) from an accredited school. Obtaining a license to practice within the United States and its possessions requires successful completion of at least 1 year of clinical training beyond the 4 years of medical school instruction. This is generally the year immediately after graduation from medical school. Training must include at least 6 months of direct patient care, both ambulatory and inpatient. Some physicians who provide care to workers and advice to employers have no specialized training beyond this, but many have undertaken additional training (residencies) in OM or another medical specialty (e.g., family practice, internal medicine, emergency medicine, or physical medicine and rehabilitation) (American College of Occupational and Environmental Medicine, 1999a).

OM is one of the four specialized areas of the American Board of Preventive Medicine (ABPM). Preventive medicine is that specialty of medical practice that focuses on the health of individuals and defined populations to protect, promote, and maintain health and well-being and prevent disease, disability, and premature death (American Board of Preventive Medicine, 1999).

In addition to knowledge common to all physicians, the following are distinct knowledge areas within preventive medicine:

- biostatistics,
- epidemiology,
- health services management and administration,
- environmental factors,
- occupational factors,
- clinical preventive medicine activities, and
- social, cultural, and behavioral influences on health.

Training in the field involves successful completion of a 2-year residency, which generally leads to board certification and which therefore closely follows the educational requirements specified by ABPM, which are as follows:

Academic year: The academic year requires the successful completion of a course of graduate academic study and the award of a Master of Public Health degree or an equivalent master's degree. The course content must include biostatistics, epidemiology, health management and administration, and environmental health.

Practicum year: The practicum year involves the planned and supervised application of the knowledge and skills acquired in the first 2 years of residency training.

- For at least 4 months the resident must engage in supervised practice within the real world of work.
- Residents must engage in collaborative work with industrial hygienists, nurses, safety professionals, and others concerned with psychosocial issues.
- Residents are encouraged to engage in research.
- Residents should not have extensive time commitments to the care of employees with minor complaints or to service functions characterized by highly repetitive or standardized procedures that do not contribute to professional growth.

Occupational Medicine Certification

Preventive medicine certification examinations are administered each year by the ABPM. The Board offers a combined examination in general preventive medicine-public health, an OM examination, and an aerospace medicine examination. To be eligible for the examinations, the physician must have been engaged in training for or practice of OM for at least 2 of the 5 years preceding the application for board certification. In addition, the candidate must meet general educational requirements for OM including 3 years of specific postgraduate medical education, that is, internship, academic year, and practicum year.

For OM, there has been an alternative pathway to certification. This alternative pathway has been limited to physicians who have graduated from a school of medicine or osteopathic medicine before January 1, 1984, but who have not formally completed all of the components described above. The following are factors considered by the board as satisfying training requirements:

Academic experience: teaching or completion of three-credit-hour, postgraduate-level academic course work in each of the four core areas of epidemiology, biostatistics, health services management and administration, and environmental health.

Practical experience: periods of full-time practice, research, or teaching in occupational health:

- Eight years is required if no other specialty certification is held.
- Six years is required if another specialty certification is held.
- Four years is required for those with a Master's of Public Health degree but without a practicum year.
- Three years is required for those with a Master's of Public Health degree and another specialty certification but without a practicum year.

All candidates judged to be acceptable by virtue of either OM residencies or equivalent training and experience must pass a 1-day written, multiple-choice test given by the Board. If the physician successfully passes the test, the Board provides a certification. Before 1998, certificates had no time limit. Effective in 1998, the certificates are valid for 10 years. Physicians who received certification in or after 1998 must take recertification tests every 10 years.

Current Status of the OM Physician Workforce

ACOEM is the professional organization for physicians interested in occupational medicine. An evaluation of the ACOEM membership shows physicians with a broad range of professional training and experience. About 40 percent of the 7,000 ACOEM members report that their primary medical specialty is OM. Family practice (8.3 percent) and internal medicine (6.8 percent) are the only other specialties designated by more than 5 percent of the members. Twenty-seven percent did not designate a specialty (Eugene Handley, ACOEM, personal communication, April 16, 1999). The primary board certifications of ACOEM members are presented in Table 2-4 (only specialties with more than 100 individuals are listed).

The percentage of ACOEM membership with board certification in OM is very low (approximately 20 percent). This may represent the lowest board certification rate of any specialty (Anstadt, 1999). Data from ABPM (C. Hyland, ABPM, personal communication, August 1999) reveals an interesting characteristic of those who are board certified. There have been 790 physicians who have obtained board certification in OM from 1992 through 1996. Approximately 55 percent obtained eligibility for board certification by the equivalency pathway. The numbers by year and by pathway are shown below:

Pathway	1992	1993	1994	1995	1996	Total	Percentage
Residency	52	68	56	79	103	358	45.3
Equivalency	60	82	88	94	108	432	54.7

These data indicate both that the specialty is growing steadily (these 790 individuals represent a significant portion of the 2,200 physicians ever certified in OM) and that the majority of today's physicians board certified in OM have joined the field without formal residency training.

A survey of a random sample of ACOEM members (The Gary Siegal Organization, Inc., 1996) provides some additional insight into the OM workforce. Despite the low percentage of ACOEM members board certified in OM, two-thirds of the survey respondents reported spending more than 90 percent of their time in OM. Although only 25 percent reported an OM clinic as their primary practice setting, three-fourths identified clinical practice as their primary activity. According to a former president of ACOEM, George Anstadt, this is a striking change from 10 years earlier. He reports that there was a dramatic decrease in the proportion of physicians working for corporations over those years, from roughly 80 percent to approximately 20 percent. The absolute number of corporate jobs has also declined, from approximately 2,400 to 1,400 (Anstadt, 1999). Preliminary data from ACOEM's 1999 Demographic Profile Survey (American College of Occupational and Environmental Medicine, 1999c) appears to confirm the trend: only 26 percent work in corporate settings, and 58

TABLE 2-4 Primary Board Certifications of ACOEM Members, April 1999

Specialty	No.	Percentage
None	1,643	30.2
Occupational medicine	1,131	20.7
Family/general practice	800	14.7
Internal medicine	779	14.3
Surgery (all subspecialties)	258	4.7
Aerospace/general preventive medicine	205	3.8
Physical medicine and rehabilitation	159	2.9
All others	469	8.6

SOURCE: Eugene Handley, American College of Occupational and Environmental Medicine, personal communication, April 1999.

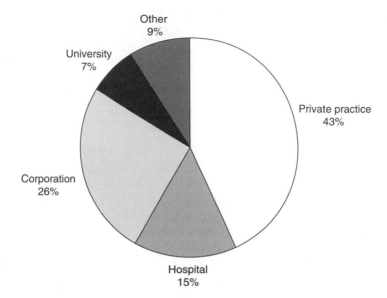

FIGURE 2-5 Primary practice setting of ACOEM members. SOURCE: American College of Occupational and Environmental Medicine (1999c).

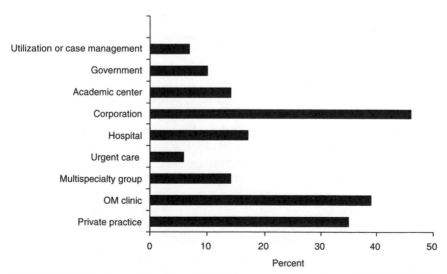

FIGURE 2-6 Practice settings for ACOEM survey respondents. SOURCE: The Gary Siegal Organization, Inc. (1996).

TABLE 2-5 Percentage of ACOEM Members Providing the Listed OM Service

OM Service	Percentage Providing Service
Acute ambulatory care	82
Medical surveillance examinations	82
Preplacement or postoffer examinations	81
Consultation for occupational or environmental problems	79
Urine collection for drug testing	78
Wellness screening	75
U.S. Department of Transportation examinations	72
Medical review officer laboratory review	70
Scheduled on-site medical	65
Disability management	61
Radiographic studies	60
Alcohol breath testing	58
Independent medical evaluations	57
Administrative oversight of occupational and environmental programs	56
Minor surgery	51
General primary care	47
Safety surveys	45
In-clinic laboratory testing	40
Industrial hygiene surveys	34
Inpatient capabilities	23
Computed tomography scanning	13
Magnetic resonance imaging	12

SOURCE: The Gary Siegal Organization, Inc. (1996).

percent now report either private practice or a hospital as their primary work setting (see Figure 2-5).

A more detailed look at where members actually spend their time is provided by asking respondents to check the settings in which they practice OM. Figure 2-6 shows the percentage of respondents indicating that they spend at least some time in the practice settings listed in the figure.

A possible explanation for the wide variety of settings indicated is provided by the answers to a question about whether the respondent's practice or group was involved in a 24-hour integrated benefits managed care plan (combined group health plan plus workers' compensation). Thirty percent answered yes to this question. Survey participants were also asked to indicate which OM services they provided (see Table 2-5).

ACOEM has several levels of membership: member, student, master, and fellow. Any physician who is engaged in occupational or environmental medicine on a full- or part-time basis or who is simply interested in providing the best possible care to working patients qualifies for ACOEM membership. Medical students, interns, and residents may apply for student membership (the number of student memberships varies at about 150).

Masters are elected from among physicians who have been ACOEM members for 3 years and meet the following qualifications:

• The applicant must have been engaged in the practice of occupational or environmental medicine on a full-time basis for 3 years.
• The applicant must meet all other requirements of the college as determined by the Board of Examiners for Fellows and Masters under the rules and procedures of the college.
• The applicant must have letters of recommendation from two fellows or masters.

Fellows are elected from among physicians who have been ACOEM members for 3 years. In addition to meeting the requirements for masters, fellows must have the following qualifications:

• The applicant must be certified in OM by ABPM or in a related specialty by another medical specialty board or must provide other documented evidence of expertise in OM acceptable to the Board of Examiners for fellow and master candidates.
• The applicant must have accumulated a total of 150 points according to a point system based on the following activities (American College of Occupational and Environmental Medicine, 1999a):

— ACOEM activities at the national level,
— component (local) society activities,
— other professional society activities,
— contributions to scientific meetings and professional literature,
— continuing education,
— faculty appointment,
— board certification.

As of July 1999, ACOEM reported a total membership of approximately 7,000. About 1,000 of these are retirees and about 150 are students, leaving about 675 active fellows, 77 active masters, and approximately 5,000 active rank-and-file members (Lanny Hardy, ACOEM, personal communication, August 5, 1999).

OCCUPATIONAL HEALTH NURSING

Occupational health nursing is the specialty practice that focuses on the promotion, prevention, and restoration of health within the context of a safe and healthy work environment. It includes the prevention of adverse health effects from occupational and environmental hazards. It provides for and implements occupational and environmental health and safety programs for workers, worker populations, and community groups. Occupational health nursing is a research-based autonomous specialty emphasizing reduction of health hazards, prevention of injury and illness, and promoting optimal health. Occupational health nurses make independent nursing judgments in providing health care and other occupational health services (Rogers, 1994).

Occupational Health Nursing History

The occurrence of disease related to occupations has been studied for centuries, with early accounts reporting the ills of exposure to the hazards of mining (see Appendix C). The emergence of occupational health nursing occurred gradually, but much of what is known about this specialty seems to have started in the late 19th century. The earliest recorded evidence of what was then called "industrial nursing" was the employment of Phillipa Flowerday by the J. & J. Coleman Company in Norwich, England, a mustard factory where she was engaged to assist the doctor in the dispensary and visit sick employees and their families in their homes (Godfrey, 1978).

In the United States, it is reported that in 1888 a group of coal mining companies hired Betty Moulder, a graduate of Philadelphia's Blockley Hospital School of Nursing, to care for ailing miners and their families (American Association of Industrial Nurses, 1976). In 1895, the Vermont Marble Company, often credited with being the first company to employ an industrial nurse, hired Ada Mayo Stuart, who provided emergency care, visited sick employees at home, taught healthy habits for living, taught child care to workers' wives, and served as the first matron of the company hospital.

The first organized industrial nursing effort began with the establishment of the industrial nurse registry in 1913, which was followed 2 years later by the establishment if the Boston Industrial Nurses Club. By 1918 more than 1,200 nurses were employed by 871 businesses. Industrial nursing continued to grow, and during the 1920s several colleges and universities offered short courses in industrial hygiene in which nurses participated. By 1930 nearly 3,200 nurses were employed in industry to provide emergency care for ill and injured workers, follow-up, and home visits.

In 1942, the American Association of Industrial Nurses (AAIN) was created for the purpose of improving industrial nursing practice and education, increasing interdisciplinary collaboration, and acting as the professional voice for industrial health nurses. At the same time, at least 15 colleges and universities offered industrial nursing courses at the baccalaureate level and an estimated 11,000 nurses were working in industry. In the 1960s, OSH became a public issue via the media and the environmental and civil rights movements. Of particular concern were mining accidents, cave-ins, and black lung disease. The importance of adequate education and training for professional disciplines was gaining increasing congressional support.

By the beginning of the 1970s, nursing education and practice were in a transitional state and more emphasis was placed on the expanded clinical role for nurses. The landmark OSHAct of 1970 provided a new stimulus and interest among both the practice and academic communities to prepare occupational health nurses and nurse practitioners at the graduate level to work in occupational health settings.

In 1977, AAIN changed its name to the American Association of Occupational Health Nurses (AAOHN), and the term "occupational health nurse" replaced the term "industrial nurse" to reflect the broad scope of practice of the occupational health nurse. In 1999, AAOHN membership exceeded 12,000, representing more than 50 percent of the occupational health nurses estimated to be practicing in that area in 1997 (Bureau of Health Professions, 1997).

The 1980s witnessed an expansion of the role of the occupational health nurse, with more involvement in health promotion, management and policy development, cost containment, research, and regulatory issues affecting practice (Babbitz, 1983; Rogers, 1988). This role continues to expand today with increased emphases on cost-effective policies and disability management.

Occupational Health Nursing Services

The occupational health nurse practices with a large degree of autonomy, but this is complemented by an interdependent role with professionals in other disciplines. Knowledge and understanding about complex work processes and related hazards, mechanisms of exposure, and control strategies that minimize or abate risks are essential to occupational health nursing practice. This requires a multidisciplinary knowledge framework guided by nursing (Rogers, 1998). Armed with interdisciplinary knowledge integrated with nursing science, occupational health nurses engage in a broad and dynamic scope of practice (Rogers, 1994):

• **Occupational Health Care and Primary Care.** In many occupational health settings, the occupational health nurse is the primary provider of service to the worker population; however, a collaborative multidisciplinary approach may be needed, depending on the problem. Primary care activities incorporate direct health care for ill and injured workers and include diagnosis, treatment, referral for medical care and follow-up, and emergency care. Nonoccupational health care may also be provided for minor health problems and chronic disease monitoring for employees with stable conditions.

• **Management and Administration.** Increasingly, the occupational health nurse is assuming a major role in the management and administration of the occupational health unit and in policy-making decisions to ensure that the OSH programs and services for workers are effective. The occupational health nurse is often the health care manager at the work site with responsibilities for program planning and goal development; budget planning and management; organizing, staffing, and coordinating the activities of the unit, including development of policy, procedures, and protocol manuals; and evaluating unit performance on the basis of achievement of goals and objectives.

• **Case Management.** The occupational health nurse acts to coordinate and manage quality health care and health care resources from the onset of an illness or injury to help return the worker to work or to an optimal alternative. Case management is often focused on high-cost, catastrophic cases; however, it is also beneficial to apply case management practices to monitoring the outcomes for every worker with an illness or injury. Early intervention is a key component of case management, as it provides for immediate problem identification and engages the worker in planning for care from the beginning of the illness or injury to recovery.

• **Health Promotion and Health Protection.** Health promotion and health protection activities are designed to improve employees' general health and well-being and to increase employees' awareness of and knowledge about toxic exposures in the workplace, lifestyle risk factors related to health and illness, and strategies that alter behaviors that are hazards to health. In addition, organizational strategies that enhance workplace health must be emphasized. Occupational health nurses practice prevention at all levels (primary, secondary, and tertiary) with an emphasis on cost containment while preserving and improving the quality of health care.

• **Counseling.** Health counseling is an integral component of occupational health nursing practice. The occupational health nurse is in the best position to provide counseling to workers, since the occupational health nurse is generally the health care provider most available to the employee. Counseling activities are intended to help employees clarify health problems and to provide for strategic interventions to deal with crisis situations and appropriate referrals. Counseling activities can relate to such areas as stress and behavioral, social, and interpersonal situations.

• **Worker and Workplace Assessment and Surveillance.** Worker health and hazard assessment and surveillance activities are designed to identify worker and workplace health problems and the state of workers' health to match the job with the employee and to protect workers from work-related health hazards. Knowledge of job demands and analysis of job tasks are essential for an accurate assessment. The occupational health nurse conducts various assessments and examinations such as an occupational history taking and preplacement, periodic, and return-to-work assessments and examinations. Preplacement examinations also help to establish baseline data for comparison with future health monitoring results. Working with other health care professionals and physicians, the occupational health nurse will want to design programs that identify vulnerable workers who are symptomatic, remove them from the exposure to prevent further insult, observe and sample the work environment to determine the exposure source(s), and reduce or eliminate the exposure agent. Use of a multidisciplinary approach can increase alternatives for problem solving, thus adding to both the effectiveness and the efficiency of programmatic interventions.

• **Community Orientation.** Community orientation activities involve the development of a network of resources that are efficiently and effectively provided to workers and employers. Collaboration and partnerships with other companies can be a vital and mutually satisfying experience that enables the occupational health nurse to develop a support system that meets the health and safety needs of employees.

• **Research and Trend Analysis.** Research activities are directed toward the identification of practice-related health problems and participation in research activities to identify factors contributing to workplace injuries and illness and ultimately to recommend corrective actions. The knowledge that is gained can then be built on to advance the profession and the practice. As part of a research team, the occupational health nurse can participate in the design, data collection, analysis, and reporting phases of research studies and can ultimately contribute to problem resolution.

• **Legal-Ethical Monitoring Activities.** Legal and ethical monitoring activities involve knowledge and integration of the laws and regulations that govern nursing practice and occupational health and recognition and resolution of ethical problems that affect workers with regard to OSH. The occupational health nurse is guided by a code of ethics that is founded on ethical theories and principles and that provides a framework regarding acts of care. The nurse needs to recognize and understand both

---

BOX 2-2
Competencies in Occupational and Environmental Nursing

- Clinical and Primary Care
- Case Management
- Workforce, Workplace, and Environmental Issues
- Regulatory/Legislative
- Management
- Health Promotion and Disease Prevention
- Occupational and Environmental Health and Safety Education and Training
- Research
- Professionalism

SOURCE: American Association of Occupational Health Nurses (1999c).

---

the personal and corporate values related to OSH and must recognize that these values may sometimes compete. The nurse is obligated to act in the best interest of the worker and provide effective leadership skills in ethical health care.

To enhance and link with the scope of practice and standards for practice, AAOHN (1999c) has recently developed competencies in occupational and environmental health nursing describing competent, proficient, and expert performances in nine categories very similar to those described in the preceding paragraphs. Competency categories are shown in Box 2-2.

Occupational Health Nursing Education

Basic nursing education is offered at the associate degree, diploma, and baccalaureate in nursing science levels. The proportion of occupational health nurses with 2-year associate degrees and diplomas (awarded by hospital-based programs of 2 or 3 years in duration that are often affiliated with a junior or senior college for the general education component of the curriculum) has decreased substantially in recent years. Figure 2-7 shows that only 44 percent of AAOHN members who responded to a recent survey reported an associate degree or diploma as their highest level of formal education (American Association of Occupational Health Nurses, 1999b). The remaining 56 percent reported attainment of a baccalaureate, master's, or doctoral degree (1 percent) as the highest level of preparation.

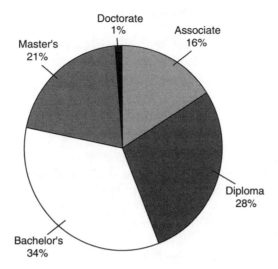

FIGURE 2-7 Highest level of formal education reported by occupational health nurses in 1999 compensation and benefits survey. SOURCE: American Association of Occupational Health Nurses (1999b).

Specialty education in occupational health nursing is generally provided at the graduate level, with both master's and doctoral degrees being offered. These programs are primarily offered through NIOSH-funded ERCs and training program grants (see Chapter 7). Master's degrees may be offered in nursing or public health. The general course content includes the following:

• nursing science, which provides the context for health care delivery, recognizing the needs of individuals, groups, and populations within the framework of prevention, health promotion, and management of care for illness or injury;
• medical science specific to treatment and management of occupational health illness and injury integrated with nursing health surveillance activities;
• occupational health sciences including

— *toxicology*, to recognize routes of exposure, examine relationships between chemical exposures in the workplace and acute and latent health effects, such as a burn or cancer, and to understand dose-response relationships;
— *industrial hygiene*, to identify and evaluate workplace hazards

so that control mechanisms can be implemented for exposure reduction;
— *safety*, to identify and control workplace injuries through the use of active safeguards and worker training and education programs about job safety; and
— *ergonomics*, to match the job to the worker with an emphasis on the worker's capabilities and minimization of the worker's limitations;

• epidemiology for study of health and illness trends and the characteristics of the worker population, investigation of work-related illness and injury episodes, and application of epidemiological methods to the analysis and interpretation of risk data to determine causal relationships;
• business and economic theories, concepts, and principles for strategic and operational planning, for awareness of the value of quality and cost-effective care, and for management of OSH programs;
• social and behavioral sciences for exploration of the influences of various environments (e.g., work, home), relationships, and lifestyle factors on worker health and determination of the interactions that affect worker health;
• environmental health for the systematic examination of interrelationships between the worker and the extended environment as a basis for the development of prevention and control strategies; and
• legal and ethical issues for ensuring compliance with regulatory mandates and contending with ethical concerns that may arise in competitive environments (Rogers, 1998).

Occupational health nurses are also prepared as occupational health nurse practitioners. That course of study provides expanded clinical training in the areas of health and disease assessment, management and treatment of illness or injury including pharmacodynamics, and occupational problem solving.

At the doctoral level, the emphasis is to prepare occupational health nursing researchers whose work may be specific to such fields of study as nursing, occupational epidemiology, environmental health, and administration of occupational health programs. As of 1998, 722 master's-level and 32 doctoral-level occupational health nurses have been added to the workplace via NIOSH training programs since their initiation in 1977 (Ann Cronin, NIOSH, personal communication, April 23, 1999).

Certification of Occupational Health Nurses

Certification in occupational health nursing is offered by the American Board for Occupational Health Nurses (ABOHN), which was established in 1972. Currently, ABOHN certifies occupational health nurses in two categories: the Certified Occupational Health Nurse (COHN) and the Certified Occupational Health Nurse Specialist (COHN-S).

ABOHN evaluates potential candidates through academic, experiential, and continuing education parameters, administers examinations, and issues certificates of qualification to those professionals who meet eligibility criteria and who pass the examination. Eligibility for the COHN examination requires licensure as a registered nurse, current employment in occupational health nursing, 75 continuing education hours over a 5-year period, and 5,000 occupational health work-related experience hours in a 5-year period. The COHN examination emphasizes the clinician, coordinator, and adviser roles. COHN-S examination eligibility has the same licensure, continuing education, and work experience requirements as those for the COHN examination, but requires an individual to have a baccalaureate degree, which must be in nursing after the year 2000, to sit for the examination. The COHN-S examination emphasizes the occupational health nurse as clinician, manager, consultant, and educator. As of April 1999, approximately 6,400 U.S. nurses hold active certification, 5,900 with the COHN-S designation and 500 with the COHN designation (Sharon Kemerer, ABOHN, personal communication, April 19, 1999).

Current Status of Occupational Health Nursing Workforce

AAOHN is the professional society for occupational health nurses. The current AAOHN membership is approximately 12,500. This represents about 50 percent of the estimated number of occupational health nurses (Bureau of Health Professions, 1997). Twenty-seven percent of the AAOHN membership indicated that they were certified in occupational health nursing, 23 percent with the COHN-S certification and 4 percent with the COHN certification (Cox, 1999).

Data from AAOHN's 1999 *Compensation and Benefits Study* (American Association of Occupational Health Nurses, 1999b) reveal that occupational health nurses have a mean of 13 years of experience in the field, with 48 percent reporting between 2 and 15 years of experience and only 7 percent reporting less than 3 years of experience. Considerations that will have an effect on the need for occupational health nurses in the future are the fact that the survey's estimate of the median age of members is 50 and the finding that 47 percent of AAOHN members are between 45 and 54 years of age.

Approximately 43 percent of AAOHN members reported being employed by various types of manufacturing industries, with "miscellaneous" industries employing the most (15 percent), followed by chemical

TABLE 2-6 Most Important Work Activities Reported by Occupational Health Nurses

Activity	Percentage Ranking Activity as Most Important	Percentage Ranking Activity Among Top Four in Importance
Primary care	44	69
Management and administration	32	58
Case management	18	65
Health promotion strategies	10	54
Health hazard assessment	8	41
Investigating injuries and illness	4	36

SOURCE: American Association of Occupational Health Nurses (1999b).

products industries (6 percent). Hospitals or medical centers employed 16 percent of members, and governments employed 8 percent. The insurance industry (5 percent) filled out the top five employers. As might be expected given these types of employers, the typical AAOHN member is employed by a relatively large company. Only 4 percent of AAOHN members are employed by companies with less than 250 employees nationwide; only 8 percent are employed by companies with less than 100 employees at their location. The member at the 50th percentile works at a company with 10,000 employees nationwide and 1,000 employees at his or her location.

Table 2-6 indicates how those who responded to the AAOHN survey reported that they spend their time at work. At the top of the list is provision of direct care (clinical diagnosis and treatment) of occupational and non-occupational illnesses or injuries, followed by case management for worker injuries and illnesses, management and administration of occupational health programs (program planning, policy development, oversight of compliance with laws and regulations), and health promotion strategies. Fewer than half of the respondents rated health assessment and surveillance of the worker or workplace or investigation, monitoring, and analysis of illnesses and injuries among their top four job functions.

The *Compensation and Benefits Survey* does not provide a comparison of data for certified and noncertified occupational health nurses, but one of the survey questions asks the respondent to select the nearest equivalent to his or her job title from a list. ABOHN asks the same question, with the same list of job titles, of all nurses taking the COHN and COHN-S examinations and provided the committee with the data for all those who

TABLE 2-7 Job Titles Reported by Two Samples of Occupational
Health Nurses

Job Title	Percentage of Random Sample of AAOHN Members	Percentage of All Certified OHN Nurses
OHN clinician	23	42
Case manager	16	4
Occupational health services coordinator	20	9
Health promotion specialist	2	1
Manager/administrator	17	20
Nurse practitioner	8	8
Corporate director	3	2
Consultant	2	6
Educator	<1	1
Researcher	<1	<1
Other	8	7

SOURCES: American Association of Occupational Health Nurses (1999b); Sharon Kemerer, ABOHN, personal communication, May 14, 1999.

are currently certified. Table 2-7 compares the AAOHN survey data for a random sample of certified and noncertified occupational health nurses with the ABOHN job title data for all certified occupational health nurses.

OTHER OSH PROFESSIONALS

In addition to the four "traditional" professions described above, thousands of other professionals contribute to OSH in U.S. workplaces. This section provides information on a few prominent disciplines whose members provide health and safety services to workers and businesses at their work sites. Many of these practitioners work on a contractual basis, or as consultants. Not included are the many health professionals (e.g., emergency medicine physicians, audiologists, and respiratory, occupational, and physical therapists) whose primary contact with OSH is provision of evaluation or treatment services to a population that includes workers.

Ergonomists

Ergonomics as a field was first defined in the 1950s by scientists in Britain who described efforts to design equipment and work tasks to fit the individual. Later, in 1957, Americans doing similar work founded the Human Factors Society, which was renamed the Human Factors and Ergonomics Society in 1992. Although the term *ergonomics* was not coined until the 1950s, scientists, mostly engineers, had been doing work in ergonomics many years earlier. The field had its origins in the military, where engineers had to design cockpits of airplanes to minimize the risk of accidents under very stressful conditions. Human factors engineers were also employed in the 1950s to design control rooms to minimize errors in nuclear power facilities. The field traditionally included design engineers, anthropometrists (who measured reach capabilities, etc.), and behavioral scientists. The main purpose was to design equipment to minimize errors and consequently, injuries. One example is the design of stairs (e.g., the heights and widths of the risers and the height of the handrail) to prevent falls on stairs. The OSH segment of the field has, in recent years, focused almost exclusively on prevention of musculoskeletal disorders such as back injuries, hand and wrist disorders (such as carpal tunnel syndrome), and shoulder and knee disorders, which often arise from a mismatch between the worker and the job that he or she must perform. The focus has also been primarily on the prevention of cumulative disorders and not injuries from trips and falls, collisions with moving objects, and other single events. This is probably due to the increasing recognition that these cumulative disorders account for a large portion of the injuries (about a third) that result in missed time from work and are the cause of even a larger portion of the workers' compensation payments. For these reasons, the primary focus of ergonomics in the past several years has been on the prevention of these cumulative trauma or work-related musculoskeletal disorders.

Ergonomist Services

An ergonomist typically is asked to evaluate a job or work task to assess the risk of musculoskeletal disorders. He or she will look at the force required (often measuring it using force gauges), how repetitious the work is (number of cycles per minute), the posture required, and other factors, such as exposure to vibration, heat, and cold and the amount of rest allowed. He or she will also help employers analyze their injury records to look for patterns that may indicate which jobs present a risk for various disorders and will then recommend modifications to the tasks or work procedures that will reduce the risk of injury from those high-risk jobs. Ergonomists are increasingly performing their work with teams of workers to obtain help from workers in identifying risky jobs and potential solutions. They are also doing extensive training of workers and managers on the principles of ergonomics and how to apply them to their work sites.

Ergonomists are often hired by industry, but only a few companies (typically, only very large companies) have in-house ergonomists. A small but growing number of ergonomists work for the government (above and beyond the human factors professionals who design equipment for the armed forces). Many teach and conduct research at universities. Most academic ergonomists also act as consultants to industry, as does the majority of ergonomists. Some consultants also supplement their practices by testifying as expert witnesses for OSHA, for injured workers in enforcement cases, in workers' compensation cases, or in third-party liability cases brought by injured individuals against manufacturers for not designing their equipment or products properly. There is also another group of professionals, like physical and occupational therapists, who are most often brought in after a worker has been injured to help redesign the worker's job so that the worker can get back to work sooner. Although they do not identify themselves as ergonomists, these professionals are also doing more and more preventive work of the sort that ergonomists do. The field has also attracted the attention of industrial engineers (the original designers of equipment and workplaces), industrial hygienists (who specialize in identifying and controlling hazards on the job), safety professionals (who have an engineering background), and OM physicians and occupational health nurses (who get involved through workers' compensation cases and who must decide on work restrictions or return-to-work orders), all of whom are now exposed to ergonomics in their professional education and training.

Education of Ergonomists

There are more than 70 graduate programs in human factors and ergonomics (a directory is available from the Human Factors and Ergonomics Society, P.O. Box 1369, Santa Monica, CA 90406). Most are in engineering schools and result in a master's or Ph.D. in industrial engineering. Only a few graduate programs (less than 30) specifically award a degree in ergonomics. The committee could not identify any undergraduate programs in ergonomics, although many programs, such as graduate and undergraduate programs in industrial hygiene, offer courses in ergonomics as part of a degree program. Ergonomists who are doing research generally have a Ph.D. in industrial engineering or, less frequently, in occupational or physical therapy or industrial hygiene. Practitioners often have had only limited course work in ergonomics and may have taken only a few short courses or continuing education courses of anywhere from 1 day to a few weeks in duration. These courses have become very popular and are an important revenue source for ergonomists. Most practicing ergonomists have acquired much of their expertise from on-the-job training.

Certification of Ergonomists

The main accreditation body in ergonomics is the Board of Certification in Professional Ergonomics (BCPE), which only began in 1990. Qualified individuals are certified by this board as Certified Professional Ergonomists or Certified Human Factors Professionals. BCPE also has lower-level certifications for entry-level professionals, the Associate Ergonomics Professional and the Certified Ergonomics Associate. At present, there are 743 Certified Professional Ergonomists/Certified Human Factors Professionals, 72 Associate Ergonomics Professionals, and 13 Certified Ergonomics Associates.

There are also competing certifications from the Oxford Research Institute: Certified Industrial Ergonomists (CIEs), Certified Associate Ergonomists (CAEs), and Certified Human Factors Engineering Professionals (CHFEPs). The Oxford Research Institute has certified about 400 people in the United States: about 180 CHFEPs, 220 CIEs, and about 30 CAEs. About equal numbers are certified outside the United States. Eighteen universities in the U.S. have been authorized to confer the Oxford Research Institute certification. The Oxford Research Institute does not require an examination like BCPE does, but the Oxford Research Institute does require that ergonomists take continuing education courses, which the BCPE does not.

The Board of Certified Safety Professionals introduced an ergonomics specialty examination for certified safety professionals in 1999.

Current Status of the Ergonomist Workforce

The Human Factors and Ergonomics Society has about 5,000 members, of which approximately 700 are students and of which about 600 are from foreign countries. However, many of these members are not "ergonomists" as they have been defined here (focusing primarily on OSH issues), but are primarily designers who deal with consumer applications. About 750 ergonomists are certified by BCPE, and another 400 are certified by the Oxford Research Institute. There are many more ergonomists in other professions, like occupational and physical therapists. For example, the American Occupational Therapy Association has a special interest group of about 1,100 members who are interested in work programs, and many of them are doing ergonomic interventions. Physical therapists are even more involved in workplace ergonomics than occupational therapy. Many got involved in ergonomics through sports rehabilitation medicine. Interest in ergonomics is also very high among industrial hygienists, although only about 250 of AIHA's 13,000-plus members claimed on their membership form that their primary responsibility was doing work in ergonomics.

A reasonable estimate of the number of people in the United States who call themselves "ergonomists" or whose primary function is to do workplace ergonomics would be about 5,000. The number is growing as ergonomics has become an important topic in the workplace and because work-related musculoskeletal disorders are such an important part of the injury picture. OSHA has already published a proposed standard for ergonomics programs, and although it has stimulated intense debate, a final rule is anticipated in 2000 or 2001. This standard is expected to spur the demand for ergonomists tremendously. However, more and more workers are being trained to combine their shop floor knowledge with limited ergonomics training to help identify and correct hazards. This trend may reduce the demand for professional ergonomists somewhat, but many companies will still be hiring ergonomists or consultants to help set up and manage their programs, at least initially. In the long run, the committee envisions an increased demand for ergonomic advice and consultation that will be met partly by full-time professional ergonomists and partly by increased training in ergonomics in the curricula of all the traditional OSH professions.

Employee Assistance Professionals

An employee assistance program (EAP) is a work-site-based program designed to assist in the identification and resolution of productivity problems associated with employees impaired by personal concerns, including, but not limited to, health, marital, family, financial, alcohol abuse, drug abuse, legal, emotional, stress, or other personal concerns that may adversely affect employee job performance. It is most often an employment benefit independent of both workers' compensation and any group health plan offered by the employer.

Employee Assistance Professional Services

The 7,000 member Employee Assistance Professional Association (EAPA) restricts full membership to persons who provide "core" employee assistance services 20 or more hours per week. These are:

• consultation with, training of, and provision of assistance to work organization leaders (managers, supervisors, and union stewards) seeking to manage the troubled employee, enhance the work environment, and improve employee job performance and provide outreach education for employees and their dependents about the availability of employee assistance sevices;

• confidential and timely problem identification and assessment for employee clients with personal concerns that may affect job performance;

• use of constructive confrontation, motivational techniques, and short-term interventions with employee clients to address problems that affect job performance;

• referral of employee clients for diagnosis, treatment, and assistance, plus case monitoring and follow-up with organizations, insurers, and other-third party payers;

• provision of assistance to work organizations in managing provider contracts, in forming and auditing relations with service providers, managed care organizations, and insurers, and in providing employee health benefits that cover medical and behavioral problems; and

• identification of the effects of employee assistance on the work organization and individual job performance.

Education of Employee Assistance Professionals

As might be expected, the individuals who provide employee assistance are typically mental health professionals. A 1998 EAPA member survey reported that 46 percent of respondents were social workers, 27 percent were alcohol or drug abuse counselors, and 12 percent were psychologists (Employee Assistance Professionals Association, 1999). None of these disciplines provide formal training in employee assistance, but all provide training in at least some of the core services listed in the previous paragraph. The working degree for EAP social workers is a master's in social work. A doctorate is the standard for psychologists. The educational requirements for alcohol and drug abuse counselors vary by state, and although states vary in their expectations, all require some combination of education, specific training in addiction, an internship, and paid counseling experience.

Certification of Employee Assistance Professionals

EAPA's certification department administers a program that provides the certified employee assistance professional designation. Necessary qualifications for sitting for the 1999 examination are

• a graduate degree in an EAP-related discipline,

- 2,000 hours of work experience in an EAP setting (over a minimum of 2 years and within 7 years of sitting for the examination),
- 15 "professional development hours" (continuing education), and
- 24 hours of advisement (supervision) by a certified employee assistance professional spread out over at least 6 months.

or

- 3,000 hours of work experience in an EAP setting (over a minimum of 2 years and within 7 years of sitting for the examination),
- 60 "professional development hours" (continuing education), and
- 24 hours of advisement (supervision) by a certified employee assistance professional spread out over at least 6 months.

Current Status of the Employee Assistance Professional Workforce

The 7,000 members of EAPA include Canadian and international members, associate members (who provide employee assistance services less than half-time), student and retiree members, and organizational members (corporations, unions, government agencies, associations, and other groups with an interest in EAP). The number of individual U.S. members is approximately 4,500. The number of certified employee assistance professionals is about 4,400.

According to the 1998 EAPA Needs Assessment Survey (Employee Assistance Professional Association, 1999), about half of the members are employed internally: 30 percent by a joint union-management arrangement, 25 percent by management alone, 8 percent by the union alone, 25 percent in an integrated model (their employer provides employee assistance both internally and to other organizations), and 10 percent in some other arrangement. The other half of the membership provides services from outside the organization: 67 percent as consultants and 33 percent as an employee of a contract EAP service provider. The committee was unable to locate any useful data on current demand for employee assistance professionals.

Occupational Health Psychologists

Occupational health psychology is an emerging specialty within psychology. In the broadest terms, occupational health psychology concerns the application of psychology to improving the quality of work life and to protecting and promoting the safety, health, and well-being of workers. The primary focus of occupational health psychology is on organizational and job-design factors that contribute to injury and illness at work, including stress-related disorders. Family and societal factors are also of interest to the extent that they influence the safety and well-being of working populations. Individual characteristics, such as skills, abilities, and temperament, and their contribution to occupational illness and injury are also subsumed under the rubric of occupational health psychology.

There are as yet no established curricula or credentials beyond a doctorate in psychology for occupational health psychology, which Quick (1999) describes as a convergence of preventive medicine and clinical and health psychology in an industrial-organizational context. The American Psychological Association and NIOSH are partners in a 5-year cooperative agreement to fund the development of graduate-level training in occupational health psychology. The purpose of this program is to develop and implement specialized graduate-level training through a course or series of courses in the area of occupational health psychology. Given that the ultimate goal is to promote occupational health psychology as a discipline within psychology, it is expected that the proposed new course(s) that is developed will be housed in the psychology department or, at a minimum, cross listed as a psychology course(s). Courses developed under this program must contain the expression "occupational health psychology" within their titles. It is anticipated that courses planned under this program will be fully developed, accredited by the university, and formally scheduled within a year of the funding date.

Examples of appropriate training activities suggested by the program announcement include (1) expansion of curricula in organizational psychology with new courses on organizational risk factors for stress, illness, and injury at work and on intervention strategies; (2) expansion of curricula and practica in clinical psychology to improve the recognition of job stress and its organizational sources; (3) expansion of curricula in human factors engineering to provide courses with more of an exclusive focus on OSH; and (4) increased exposure of behavioral scientists to research methods and practice in public and occupational health and epidemiology.

Programs at six universities were funded in 1998 and 1999, and additional applications are expected by May 2000. The six programs are at various stages of development, but descriptions of their proposed efforts and academic partners suggest that the field is likely to develop along the lines of industrial-organizational psychology, whose practitioners in the Society for Industrial and Organizational Psychology define themselves as

• scientists who derive principles of individual, group, and organizational behavior through research;

- consultants and staff psychologists who develop scientific knowledge and apply it to the solution of problems at work; and
- teachers who train students in the research and application of industrial-organizational psychology.

A recent membership survey by the Society for Industrial and Organizational Psychology (Burnfield and Medsker, 1999) found that 34 percent of its members are employed in academia, 31 percent are consultants or self-employed individuals, 16 percent work for organizations in the private sector, and 7 percent are employed in the public sector.

SUPPLY, DEMAND, AND NEED

Supply

This chapter began by noting that without an extensive survey it would be impossible to describe the full spectrum of individuals who contribute to OSH programs in U.S. workplaces. In the absence of such a survey the committee relied on membership in the major OSH professional organizations and certification by appropriate professional boards as estimates of the current supply of OSH professionals. Table 2-8 summarizes the current OSH professional workforce described in this chapter, that is, safety professionals, industrial hygienists, OM physicians, occupational health nurses, ergonomists, and employee assistance professionals. None of those professional organizations claim to have as members 100 percent of those who are eligible, and it is doubtful that any of

TABLE 2-8 Estimated Number of Active OSH Professionals in the United States, 1999

Type of Professional	No. of Professional Association Members	No. of Certified Individuals
Safety professionals	33,000	10,000
Industrial hygienists	14,000[a]	6,400
Occupational medicine physicians	7,000	1,150
Occupational health nurses	12,500	6,400
Ergonomists	5,000[b]	1,000[b]
Employee assistance professionals	4,500	4,400
Total	76,000[a]	28,950

[a]Total includes AIHA membership plus ACGIH membership minus duplicates.
[b]Total includes an unknown number of industrial hygienists and safety professionals.

the organizations even have as members 100 percent of those certified in their field. The numbers in Table 2-8 are therefore a very conservative estimate of the OSH professional workforce and are very likely a gross underestimate. Some measure of the extent of the underestimate may be taken from the findings of a 1996 national survey (Bureau of Health Professions, 1997) of registered nurses. This survey of more than 25,000 nurses found that 1.0 percent of respondents reported occupational health as their primary employment setting, leading to an estimate of 21,575 nurses working in the field nationwide, about 73 percent higher than the AAOHN membership total. The committee had no way of estimating the undercount for the other OSH professions, but if the proportion is similar to that for occupational health nurses, the total of 76,000 in Table 2-8 might well be as much as 50,000 short as an estimate of the size of the current OSH professional workforce.

Demand

Demand for OSH personnel, that is, employment opportunities, is equally difficult to estimate without an extensive survey of current and potential employers. The committee had neither the resources nor the license for such a survey and could find no evidence for a recent survey of this sort in the published literature. Anecdotal evidence, average salaries reported by the professional societies, computer modeling of industrial hygiene positions, and an informal survey of NIOSH-supported training programs by the committee suggest that overall supply appears to be roughly consonant with market demand. It seems possible, if not likely, that this is due in large measure to the elasticity of employer demand, that is, a willingness to accept less educated or less experienced professionals rather than pay a premium for the most highly qualified individuals. Such elasticity may also partially explain the relatively low percentage of OSH professionals who are board-certified in their field (Table 2-8). Certification is certainly not synonymous with expertise, but it does serve as a notice to prospective employers that the holder has been judged competent by his or her peers. Employers are apparently not willing to pay a sufficient premium for this guarantee to induce the majority of OSH professionals to gain certification. Conversely, the fact that certification in any of the professions does command some additional pay and benefits might suggest that the market is indeed calling for more certified personnel. It is quite possible, however, that additional certified professionals would simply displace noncertified OSH personnel. That is, the number of positions for OSH professionals (demand) might well remain unchanged.

Need

None of the earlier discussion should be taken to suggest that there are no unmet needs in the field. By this the committee means shortfalls or other deficiencies in the current OSH workforce that the committee believes ought to be corrected or ameliorated for workers to be protected to the extent that current knowledge allows. The continuing annual reports of 6,000 fatal workplace injuries and 6 million nonfatal workplace injuries and the estimated 60,000 annual deaths from occupational illness (Leigh et al., 1997) are ample evidence that the workforce needs more protection than its employers are providing.

An earlier section of this chapter reports that although 3 percent of safety professionals in the American Society for Safety Engineers have doctorates, only nine U.S. universities offer a doctoral degree in safety, and the committee was able to identify only one dissertation since 1995 that focused on the traditional domain of safety professionals: prevention of sudden traumatic injury. This is discussed further in Chapter 7, along with a suggestion for action, but it should be clear that the current low level of doctoral graduates is not sufficient to maintain the faculty presently training the bachelor's- and master's-level safety workforce.

A similar situation exists in OM, in which the small number of board-certified OM specialists means that injured or ill workers must often obtain care from physicians who are not specialists in the area. A 1988 Institute of Medicine report that explored the barriers that are keeping primary care physicians from competently meeting the needs of patients with environmental and occupational problems pointed to a lack of specialty-trained physicians, that is, board-certified OM physicians, to serve as educators and consultants. Subsequent publications (Castorina and Rosenstock, 1990; Institute of Medicine, 1991) estimated this shortfall to be 3,100 to 5,500 positions. Although the authors suggested that 1,500 to 2,000 of those positions might require only primary care practitioners with "special competence" in occupational and environmental medicine (i.e., additional training but not a residency), they pointed out that closing that gap would require increases in the number of individuals undergoing graduate specialty training by a factor of three to five for a period of 10 years (as well as significant changes in the structure and funding of universities and public health departments). Ten years later, the number of board-certified OM specialists remains unchanged. Chapter 7 explores some non-monetary reasons for the persisting gap and provides a suggestion for a means of reducing it.

The major shortfall in the field of occupational health nursing is similar to that in OM: not so much a shortage of practitioners as a shortage of practitioners with formal training in the field. In the case of occupational health nurses this is a master's degree. Because no residency is involved, a doubling of the annual number of master's-level occupational health nursing graduates is not an unrealistic goal if established professionals are not required to abandon an ongoing career for a year or more. Chapter 8 explores some emerging means of accomplishing this.

The distinction between demand and need is nowhere more aptly illustrated than by examination of the makeup of the U.S. workforce as a whole. Some 56 percent of U.S. workers are employed by firms with less than 100 employees (National Institute for Occupational Safety and Health, 1999). The majority of the traditional OSH professionals, however, are employed by midsize to large businesses and government agencies. Small businesses as a group apparently believe that they do not need OSH professionals or cannot afford them. Most workers will thus seldom, if ever, encounter one of these OSH professionals. Even when injured, they may receive treatment in emergency rooms or ambulatory clinics where the treating physicians and nurses have neither the time nor the training to deal with issues of causation and prevention of the injury-producing event.

The training of traditional OSH professionals is considerably simplified by the nature of their practice in midsize and large industries and government agencies. To a large degree the framework is constructed around the regulatory system and the workers' compensation insurance system. It is a top-down system that, for those segments of the economy, reaches all the way to the worker at risk. This part of the system includes professionals who operate from consulting practices and who play an increasing role in midsized industries. It is another matter entirely to address the development and the training of OSH personnel to deal with that large majority of the workforce who have no routine access to the OSH system. The few OSH professionals who do focus on small businesses and workplaces are likely to work for government or public interest groups. For these professionals, media and communication skills are likely to be the most important requirement. For example, the "right-to-know" concept was popular in the 1980s occupational and environmental health community as a means of stimulating a bottom-up demand (i.e., turning a need into a demand) for a safer and healthier environment among those directly affected. Putting that approach to work in the workplace, however, continues to be a challenge and, on the prevention side at least, the OSH system in the United States principally affects that portion of the workforce that is employed in large industry or by government or that is represented by organized labor. As subsequent chapters will elaborate, the workplace of the future will increasingly be dominated by small service-producing businesses that are widely distributed and that utilize an increasingly diverse and transient workforce. What is most needed now and will be needed even more in the coming decades, in addition to the traditional OSH professionals, is a new and different model of practice, perhaps one that even uses new categories of OSH personnel created by training managers, supervisors, and workers already employed in these small workplaces.

3

The Changing Workforce

ABSTRACT. Projected changes in the labor force of the next decade will result in a workforce with a larger proportion of workers over age 55 and larger proportions of women, Blacks, Hispanics, and Asians. Women and older workers have lower injury and illness rates than the labor force as a whole, although injured older workers take longer to return to work. The Americans with Disabilities Act of 1990 mandated reasonable accommodation for workers with a disability, and the number of employed persons with a severe disability grew tremendously in the 1990s. The committee concludes that all aspiring occupational safety and health (OSH) professionals must be made aware of ethnic and cultural differences that may affect implementation of OSH programs. In addition, the committee believes that OSH programs are social as well as scientific endeavors and that health care disciplines and professionals should reflect the social makeup and the diversity of thought and experience of the societies they serve, so it will be important that members of all racial and ethnic groups be actively recruited. In addition, education in all OSH professions will need to include instruction on changes in the physical and cognitive abilities of older workers, the interaction of disabilities and chronic diseases with workplace demands, and communications skills to reach minority workers, workers with low levels of literacy, and those for whom English is a second language. A knowledge of and willingness to work with mass media may be required to reach workers at home as well as at work.

Many of the changes expected in the U.S. workforce in the next decade will be continuations of trends already well under way. For that reason, the committee's analysis of future changes likely to affect the training of OSH personnel begins with a review of prominent demographic trends of the last few decades. Among the significant trends discussed in this chapter are the rapid growth in the number of Hispanics and Asians in the labor force and the continued increase in women's share of the workforce. The aging of the baby boom generation has also increased the number of older workers in the labor force.

The primary source of data for this chapter is the Bureau of Labor Statistics, especially its periodic data collections such as the Current Population Survey. Many of these data are directly accessible at the Bureau of Labor Statistics website (http://www.bls.gov/oshhome.htm), but many of the tables and figures are the result of a specific request and can be replicated only by contacting the Bureau of Labor Statistics and asking for a tabulation of the specific data in question.

REVIEW OF PAST WORKFORCE CHANGES

Several developments have been important contributors to the labor force changes of the past few decades. Among such factors are the completion of the entry of the baby boom generation into the workforce and the impact of the smaller birth cohort that followed the baby boomers into the workforce. In addition, the continued entry of women into the labor force over the past few decades has profoundly affected the distribution of men and women in the workforce. The third important workforce development over the past few decades has been the increased immigration of Asians and Hispanics into the United States and their subsequent entry into the workforce.

A closer look at each of these elements of labor force change should provide a better understanding of likely future labor force developments. Overall, the workforce growth in the most recent 10-year period (1988 to 1998) has been both in numerical and in percentage terms slower than that in the previous decade (an increase of 16 million [10 percent] workers in the period from 1988 to 1998 compared with an increase of 19 million [12 percent] from 1978 to 1988). This slowing in the rate of growth of the labor force reflects the much smaller size of the birth cohort that followed the baby boomers into the workforce. Labor force change has also been influenced by the fact that although the number of women in the workforce is still growing faster than the number of men, the gap between the growth rates of these two groups has narrowed appreciably.

Age

The significant difference in the sizes of the two birth cohorts—the baby boomers and the group that followed them—has had and will continue to have a pronounced effect on the age distribution of the workforce. Over the past decade, for example, there has been an absolute decline in the number of workers in the labor force under age 25 (Figure 3-1). Conversely, the most significant workforce growth during the 1990s has been among workers 25 to 54 years of age, reflecting the aging of the baby boom generation.

Gender

Women's share of the workforce has been increasing for several decades as their pattern of labor force activity more and more mirrors that of men. Thus, women increasingly enter the labor force at a young age and become permanent labor force participants. As a consequence of this long-range change, women's share of the labor force has been growing steadily. However, in the most recent decade growth has slowed, reflecting the fact that the women's share of the labor force has already reached significant proportions. Women's share of the labor force had by 1998 grown to nearly 46 percent.

A related change in women's labor force participation has been that, increasingly, women with children have been working. Women were formerly well represented in the labor force before they had children or after their children had completed school. That is no longer true, as shown in Table 3-1, which indicates that the labor force participation of women

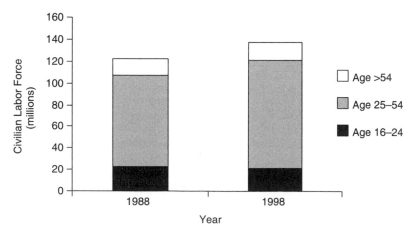

FIGURE 3-1 Age of civilian labor force, 1988 and 1998 (see Table 3-4 for sources).

TABLE 3-1 Percentage of Mothers in Labor Force, 1975–1998, by Age of Youngest Child* in March 1973

Year	All Mothers	Mothers with:		
		Children Ages 6–17	Children Ages 3–5	Children Aged < 3
1975	47.3	54.8	44.9	34.1
1980	56.6	64.3	46.8	41.9
1985	62.1	69.9	59.5	49.5
1990	66.7	74.7	65.3	53.6
1995	69.7	76.4	67.1	58.7
1996	70.2	77.2	66.9	59.0
1997	72.1	78.1	69.3	61.8
1998	72.3	78.4	69.3	62.2

*Children are a woman's own children and include sons, daughters, adopted children, and stepchildren. Not included are nieces, nephews, grandchildren, and other related and unrelated children.

SOURCES: Bureau of Labor Statistics, Current Population Survey, March Supplements, selected years.

with children under age 3 reached 62.2 percent by 1998. Women's labor force participation patterns have also changed in another way in that in the past women tended to withdraw from the labor force for childbearing; that is no longer the pattern. Consequently, there was a distinct drop in women's labor force participation rate during years of peak childbearing. That is no longer the case.

Race and Ethnicity

The composition of the workforce by race and ethnicity has also undergone profound changes over the past few decades. These changes have been the result of increased immigration, particularly among Asians and Hispanics, and of the higher birth rate among Hispanics. Because of the rapid changes in the proportion of other groups in the workforce, white non-Hispanics' share of the workforce, which was 74 percent in 1998, had shown a decline of over 5 percent since the mid-1980s. Although their share of the workforce has been declining, the number of white non-Hispanics in the labor force is still growing, but the rates of growth are slower than those of Blacks, Hispanics, and Asians and Others.[1] The share of the last two groups in the workforce in particular has

[1]Bureau of Labor Statistics categories. The committee uses the term African-American whenever possible because that is the term preferred by members of that group, but uses

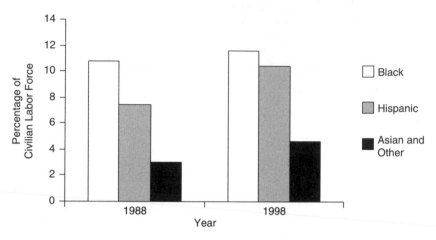

FIGURE 3-2 Minorities in the civilian labor force, 1988 and 1998 (see Table 3-5 for sources).

shown very rapid growth. Hispanics' share of the labor force by 1998 had grown more than 14 percent and was rapidly approaching Blacks' share of the workforce (Figure 3-2) (see also Table 3-5 later in this chapter).

Disability

One factor that could be important to the training needs of OSH personnel in the future is the disability status of the population. The Bureau of the Census, in its periodic survey of income and program participation, provides a brief review of the disability status of the population and the impact that this has on employment. The range of disability included in the survey and the level of employment of each of these groups are shown (Table 3-2) for two periods of time. Not only is the employment of those with disabilities lower but their employment is also significantly affected by the severity of their disability. Furthermore, although the employment rates of both groups increased from 1991–1992 to 1994–1995, the rate of increase was faster for those with disabilities and was fastest for those whose disabilities were most severe (a 27 percent increase in the number employed over this 3-year period). These changes came during a period when the implementation of the Americans with Disabilities Act was

---

the term "Blacks" when discussing Bureau of Labor Statistics data. The "Asian and Other" labor force group includes Asians, Pacific Islanders, Alaskan Natives, and Native Americans.

TABLE 3-2 Employment of Persons Ages 21–64 Years, 1991–1992 and 1994–1995

Group	1991–1992		1994–1995	
	Number	Percent Employed	Number	Percent Employed
All persons	144,778	75.1	149,396	76.2
With no disability	117,194	80.5	119,902	82.1
With any disability*	27,584	52.0	29,467	52.4
Severe	12,568	23.2	14,219	26.1
Not severe	15,016	76.0	15,248	76.9
With a mental disability			6,012	41.3
Uses a wheelchair	495	18.4	685	22.0
Does not use a wheelchair, has a cane, crutches, or a walker for ≥6 months	1,108	17.6	1,609	27.5
Unable to perform one or more functional acts			6,841	32.2
Unable to see words or letters	563	25.6	568	30.8
Unable to hear normal conversation	324	58.2	358	59.7
Unable to have speech understood	122	24.4	119	27.7
Unable to lift/carry 4.5 kg (10 lbs.)	3,028	22.3	3,017	27.0
Unable to climb stairs without resting	3,516	20.5	3,736	25.5
Unable to walk 3 city blocks	3,182	20.8	3,547	22.5

*In this survey a person was considered to have a disability if the person met any of the following criteria: (1) used a wheelchair, (2) had used a cane or similar aid for 6 months or longer, (3) had difficulty with a functional activity such as seeing or hearing, (4) had difficulty with an activity of daily living such as bathing or dressing, (5) had difficulty with an activity of daily living such as going outside the home or using the telephone, or (6) was identified as developmentally disabled or as having a mental or emotional disability. Those aged 16 to 64 were considered disabled if they had a condition that limited the kind and the amount of work that they could do.

SOURCE: Bureau of the Census, based on the Survey of Income and Program Participation.

taking place, but it is not possible to separate the effects of this act from the effects of a favorable economic climate in the 1990s that improved employment opportunities for all groups. Particularly during the last decade, the favorable economic and employment climates have benefited not only those with disabilities but also immigrants and low-skilled individuals in the labor force.

Literacy

The National Center for Educational Statistics of the U.S. Department of Education in 1992 conducted the National Adult Literacy Survey (National Center for Educational Statistics, 1993). That survey revealed that 40 million to 44 million adult Americans (21 to 23 percent of the adult population) fell in the lowest level of reading proficiency. Many adults with this level of literacy could perform only simple, routine tasks involving brief and uncomplicated texts and documents. Others were unable to perform even these types of tasks, and some had such limited skills that they were unable to respond to the literacy survey. A number of factors explained why so many adults demonstrated English literacy skills at the lowest proficiency level. Twenty-five percent of those in the survey who performed at the lowest level were immigrants and may have just been learning English. Nearly two-thirds of those who performed at this level had terminated their education before the completion of high school. One-third were older than age 65, and more than one-quarter had physical, mental, or health conditions that kept them from participating fully in work, school, or other activities. The results of the literacy survey are very important to determining the training needs for future OSH personnel, who must be trained to deal with workers with all levels of literacy, particularly given that labor force trends predict continued growth in the labor force participation of groups such as immigrants from which workers with low levels of literacy are more likely to be drawn. Research (Chiswick and Miller, 1998) has demonstrated that language proficiency is related to employability and income: language proficiency is greater among immigrants with higher levels of schooling, longer duration in the United States, and younger age of immigration and is lower among those who live among others who speak only their native language.

Economic Trends

Before turning to an examination of future labor force developments, it may be helpful to briefly review economic trends. Economic trends influence labor force developments, and these in turn are important determinants of future economic developments. One measure is real disposable income per capita. This is a measure of the income available on average to everyone in the population after adjustment for inflation and taking into account the effect of taxes. It is one measure that is used to show changes in the standard of living over time.

Table 3-3 shows that disposable income grew steadily over the period from 1986 to 1998, and the economic expansion underlying that growth no doubt played a significant role in some of the employment trends

TABLE 3-3 Real Disposable Per Capita Income, Selected Years, 1986–1998, and Projected to 2010

Year	Per Capita Income
1986	$16,939
1988	$17,650
1996	$18,989
1998	$19,790
2010	$23,120

SOURCES: Office of the President (1999) and Bureau of Economic Analysis (1999). Projections to 2010 are by the Committee to Assess Training Needs for Occupational Safety and Health Personnel in the United States, Institute of Medicine, using the Bureau of Labor Statistics' projected trends for 1996 to 2006.

discussed above. The rate of employment of minorities, for example, has often been hit harder by recession than the rate of employment of the majority population, and there may be less demand for older workers in a less robust economy.

A LOOK TO THE FUTURE

What will be the labor force changes in the future, particularly in the period to 2010? In many ways this period will be affected by the same population changes that have had profound influences on the labor forces of recent decades. Discussing the labor force of 2010 requires only a relatively modest level of speculation, and the committee believed that 2010 was sufficiently distant to give some sense of the labor force of the future but not so far in the future that the potentially profound impact of unforeseen technologies would be overly influential.

Overall Changes

The growth of the labor force will continue to increase, but at a slower rate of growth, reflecting the fact that the new entrants into the workforce will be drawn from smaller birth cohorts than those from which the labor force was drawn from earlier. In 1998 the labor force had reached more than 137 million. By 2010 that number should reach more than 155 million, but this increase of 18 million would reflect a growth of only slightly over 1.0 percent a year—or about one-half the rate of workforce growth achieved in the late 1980s and early 1990s. This slowdown in the rate of growth reflects both the smaller size of the birth cohorts reaching labor force age and the fact that such a large share of women are already in the workforce that they no longer constitute the major source of new entrants to the labor force that they have been over the last two decades (see Table 3-5).

Age

The number of people in the youngest group in the labor force (those 16 to 24 years of age), which had declined in the 1990s, should begin to increase after 2000, although only modestly. This reflects the fact that the children of the baby boom generation make up a somewhat larger birth cohort than the group slightly older than them. Because of this slightly older but smaller birth cohort, absolute declines in the number of people in the labor force aged 25 to 44 years are expected to take place in the decade ahead. The age group with the most rapid growth in the next decade will be that consisting of people aged 45 to 54—the baby boom generation—the oldest of whom will be nearing retirement age by 2010.

Another important question is the longer-term prospects for labor force participation of those age 65 years or older; that is, has there been a change in the average age of retirement? Evidence is clear that the long-term trend toward earlier retirement has stopped. Evidence for a reversal, however, is mixed: the 1994 labor force survey that found more older workers was redesigned, and the subsequent 4 years of data are not enough to signal a clear long-term trend.

Gender

Women's share of the workforce, as noted earlier, has grown steadily and has already reached a high level. Therefore, even though the number of women in the workforce is expected to grow slightly faster than the number of men, women's share of the labor force is expected to increase only from 46 percent in 1998 to almost 48 percent by 2010 (Table 3-4).

Race and Ethnicity

Although the rate of growth of participation in the labor force by all racial and ethnic groups is expected to slow in the decade ahead, the pattern of more rapid growth for minority groups than for non-Hispanic whites is expected to continue. In particular, rapid rates of growth are projected for the Hispanic and the Asian and Other groups. The number of Hispanics in the workforce should be equal to the number of Blacks

TABLE 3-4 Civilian Labor Force, 1988, 1998, and Projected to 2010*

Group	1988 No. (millions)	Percent	1998 No. (millions)	Percent	Projected 2010 No. (millions)	Percent
TOTAL	121.7	100.0	137.7	100.0	155.4	100.0
Men	66.9	55.0	74.0	53.7	80.9	52.1
Women	54.7	45.0	63.7	46.3	74.5	47.9
16 to 24 years old	22.5	18.5	21.9	15.9	25.7	16.5
25 to 54 years old	84.0	68.0	98.7	71.7	103.3	66.5
>55 years old	15.1	12.4	17.1	12.4	26.4	17.0
White	104.8	86.1	115.4	83.8	128.0	82.4
Black	13.2	10.8	16.0	11.6	18.2	11.7
Asian and other	3.7	3.0	6.3	4.6	9.2	5.9
Hispanic origin	9.0	7.4	14.3	10.4	19.5	12.5
Other than Hispanic	112.7	92.6	123.4	89.6	135.9	87.5
White, non-Hispanic	96.1	79.0	101.8	73.9	110.4	71.0

*The civilian labor force includes all employed individuals in the economy (except the uniformed military service) and all who are actively seeking employment (or the unemployed). The projections include estimates for the undercount of the population in the Census of Population and also include estimates of future immigration, both documented and undocumented.

SOURCES: Historical data are from the Bureau of Labor Statistics Current Population Survey, annual averages, selected years. Projections are an extension of the Bureau of Labor Statistics projections for 1996 to 2006 (Bureau of Labor Statistics, 1997) to 2010 by the Committee to Assess Training Needs for Occupational Safety and Health Personnel in the United States, Institute of Medicine.

sometime in the middle of the first decade of the 21st century. Although the number of Blacks in the workforce will be growing slightly faster than the average, their share of the workforce can be expected to increase only moderately above their current level of 11.5 percent. The Asian and Other group is expected to be the fastest growing group in the labor force, and its share should approach 6 percent by 2010.

Labor Force Entrants and Leavers

Important insights into future labor force developments can be seen by using the facts presented above to calculate who will be entering the labor force and who will be leaving the labor force in the first decade of the 21st century. As noted earlier, the labor force is expected to grow by nearly 18 million during the period from 1998 to 2010. However, that increase is the net result of more than 47 million who will be entering the workforce and nearly 30 million who are expected to leave the workforce over the same period (Table 3-5). Importantly, the gender and the racial and ethnic compositions of those coming into the labor force and those who are expected to leave the labor force are much different. To illustrate, of those who are expected to leave the workforce during the period from 1998 to 2010, 57 percent are men and 43 percent are women. Entrants, on the other hand, are expected to be nearly equally divided between men and women. Furthermore, the Hispanic and the Asians and Other groups represent nearly one-quarter of the entrants expected over the period to 2010 but make up only about 11 percent of those who will be leaving the labor force. If Blacks are included with the other two minority groups, more than 40 percent of the entrants to the workforce from 1998 to 2010 are projected to be Black, Hispanic, or Asian and Other (Figure 3-3).

Disability

It would be desirable to have data on the employment prospects for those with disabilities in 2010, but no such projections have been prepared. The levels of employment of persons with disabilities have increased over the last decade, and that improvement in the employment prospects for those with disabilities will probably continue. However, as noted earlier, at least part of that improvement is attributable to the favorable economic climate of the 1990s. Consequently, a less favorable economic climate could slow that trend in the future.

IMPLICATIONS OF CHANGING DEMOGRAPHICS FOR OCCUPATIONAL INJURIES AND ILLNESSES

The labor force changes projected for the next decade imply a workforce with a larger share of workers over age 55, a slightly larger share of women, and a rapidly growing share of Blacks, Hispanics, and Asians. The rate of employment of persons with disabilities has risen sharply since passage of the Americans with Disabilities Act in 1990 and may well continue through the next decade. What, if any, implications do these trends hold for occupational injuries and illnesses and, ultimately, for the training needs of those who deliver OSH services?

In one important dimension of occupational injury and illness from 1993 to 1996, women incurred less than 1/10 of the job-related injuries and about 1/3 of the nonfatal injuries and illnesses that required time off from work (Bureau of Labor Statistics, 1998b). Of the 32,000 job-related fatalities that occurred from 1993 to 1996, only 8 percent occurred among

TABLE 3-5 Total Civilian Labor Force, Entrants, Leavers, and Stayers, 1988, 1998, and Projected to 2010 (number, in millions)

Group	1988 Total Force	Entrants	Leavers	Stayers	1998 Total Force	Entrants	Leavers	Stayers	2010 Total Force
Total	121.6	35.0	19.0	102.6	143.7	47.5	29.8	107.9	155.4
Men	66.9	18.4	11.4	55.6	74.0	23.8	16.9	54.1	80.9
Women	54.7	16.6	7.7	47.1	69.7	23.7	12.9	50.8	74.5
White*	96.1	24.4	15.8	80.3	101.8	28.0	20.4	81.4	110.4
Black*	13.0	4.6	2.0	11.0	15.6	7.4	6.1	9.5	16.9
Hispanic	9.0	6.3	0.9	8.1	14.3	7.1	1.5	12.8	19.5
Asian*	3.6	2.8	0.3	3.2	6.0	5.0	1.8	4.2	9.2

*Each of these is the non-Hispanic portion of this labor force group.

SOURCES: Historical data are from the Bureau of Labor Statistics Current Population Survey, annual averages, selected years. The projections to 2010 are by the Committee to Assess Training Needs for Occupational Safety and Health Personnel in the United States, Institute of Medicine, based on the Bureau of Labor Statistics projections for 1996 to 2006.

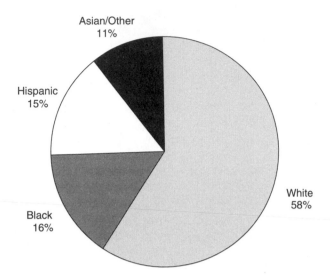

FIGURE 3-3 New entrants to the labor force projected from 1998 to 2010, by race and ethnicity, as a percentage of total new entrants (see Table 3-5 for sources).

women, even though they represented nearly 50 percent of the workforce. However, women were more likely than men to be injured in an incident of workplace violence; although women were the victims in only about one-third of the nearly 2 million annual incidents of workplace violence reported by the U.S. Department of Justice (1998), 65 percent of the nearly 23,000 reported assault-related injuries were incurred by women. Assaults are also the leading cause of workplace fatalities for women. About 70 percent of the nonfatal assaults resulted in lost work and occurred in nursing homes, in hospitals, or in the course of social service, all of which are industries that are projected to grow very rapidly. The increase in women's share of the workforce thus suggests the possibility of declines in the numbers of injuries and illnesses, albeit with the important exception of workplace violence. However, as gender barriers continue to erode over time and the distribution of jobs between men and women becomes more similar, the distribution of fatalities, injuries, and illnesses between men and women may also become more similar.

A second implication for the OSH community lies in the fact that the 10 million Americans ages 55 and over who work for wages and salaries in the private sector are one-third less likely than younger workers to be hurt seriously enough to lose time from work. However, when they are injured seriously enough to lose time from work, older workers typically require 2 weeks to recover before returning to work, twice the recuperation time that younger workers need—a median of 10 days for those involving workers over age 55 compared with a median of 4 or 5 days for workers under age 35. The reasons for this are that workers over age 55 both sustain more especially disabling injuries than younger workers and take longer to recuperate from similar injuries (Bureau of Labor Statistics, 1996). Since the number of workers in the over age 65 age group is projected to grow as a share of the workforce as the aging baby boomers move into it in increasing numbers, there are likely to be fewer injuries and illnesses, but the injuries and illnesses will be more severe ones and will require longer recuperation times. OSH professionals involved in case management and back-to-work programs need to be aware of this difference, which is one example why age-specific information on physical and cognitive capabilities should be included in the future training of OSH professionals.

In regard to the other demographic trend, that is, that Blacks, Hispanics, and Asians will make up a larger share of the workforce, the types of data needed to make a reasonable estimate of the effects of that trend on workplace injuries and illnesses do not exist. Although data on the number of injuries and illnesses by race and ethnicity are available, the information necessary to calculate the rates of injuries and illnesses—namely, hours worked by race by industry and occupation—are not available. It is safe to say, however, that any factor, be it cultural assumptions and habits or low levels of comprehension of the English language, that impedes communication of health and safety information is likely to increase the number of workplace injuries and illnesses.

IMPLICATIONS FOR TRAINING NEEDS OF OCCUPATIONAL SAFETY AND HEALTH PERSONNEL

The changing demographic makeup of the workforce is an important element in determining the training needs of future OSH personnel. Clearly, the racial and ethnic compositions of future entrants to the workforce will be different from those who are currently in the labor force. Trainers must be cognizant of the fact that new workforce entrants are more likely to be members of a minority group, women, and immigrants with various levels of English proficiency and with low levels of literacy. Also, the minority groups entering the labor force tend not to be distributed uniformly across the country. These changes are important to both the education and the recruitment of future OSH personnel. All aspiring OSH professionals must be made aware of ethnic and cultural differences that may affect implementation of OSH programs (for example, distrust of health care professionals). In addition, the committee believes that OSH programs are social as well as scientific endeavors and that health care disciplines and professionals should reflect the social makeup and the diversity of thought and experience of the societies they serve. It will therefore be important that members of all racial and ethnic groups be actively recruited, and that geographic areas where rapidly growing minority groups are having a major impact on the labor force have substantial numbers of minority OSH personnel. None of the professional societies that the committee relied on for demographic information collected data on race, but committee members who regularly attend OSH professional meetings reported that an observer could not fail to note the paucity of persons of color. In addition, even though the employment rate for those with disabilities has been improving, it is still significantly below the rate for those who have no disabilities. Continuing efforts may be necessary to sustain and enhance these improvements. The fact that a sizable portion of the adult population has a low level of literacy poses important issues as well.

The changes discussed in this chapter seem to call for new or significantly different training directed at the worker as well as the workplace. Certainly, OSH professionals will increasingly need to be familiar with the aging process and the interaction of disabilities and chronic diseases with workplace demands. In addition, the training must emphasize communications skills so that OSH personnel can reach workers with low levels of literacy and those for whom English is a second language.

4

The Changing Workplace

ABSTRACT. The industrial and the occupational structure of the U.S. economy changed in important ways during the decade ending in 1998. Among goods-producing sectors, only construction added jobs and the manufacturing and mining sectors both lost jobs. The service-producing sector, on the other hand, led by retail trade and business and health services, grew dramatically. The rate of growth in the numbers of individuals in four occupational groups—executive, professional, technician, and service—is projected to be more rapid than the overall rate of growth in the economy, and, consequently, their share of the workforce will increase as this trend continues. Although there are some important exceptions, the rate of occupational injuries has been higher in the declining industries such as manufacturing than in industries that are expected to continue to grow, such as retail trade. The majority of workers are now employed by small firms, and that will be increasingly true for the new jobs being created. More work will be contracted, outsourced, and part-time. Substantial numbers of workers will hold multiple jobs and will change jobs more frequently. It is anticipated that an increasing number of workers will work at home, and in some sectors there has been a decline in the number of workers represented by unions.

The committee concludes that these changes, as a whole, describe a workplace very different from the large fixed-site manufacturing plants in which occupational safety and health professionals have made the greatest strides. The changes complicate the delivery of occupational safety and health services and argue for types of training and delivery systems that are different from those that have been relied upon to date. Simply increasing the numbers or modifying the training of occupational safety and health (OSH) professionals will not be sufficient, since the primary difficulty will be access to either underserved workers or under-

served workplaces. Extensive new regulation is possible but seems unlikely. Other problems not susceptible to site- or group-specific interventions (smoking, seat belt use, and drunk driving) have been attacked
with broad public education campaigns. The committee calls for systematic exploration of new models for implementing occupational health and
safety programs for the full spectrum of U.S. workers.

Just as the U.S. workforce has been changing steadily in the last two
to three decades, the U.S. workplace has been undergoing even more
dramatic changes as the country moves away from heavy industry into
the information age. Therefore, this chapter begins as the previous one
did, with a short review of changes that have occurred in the recent past.
As in the previous chapter, the primary source of data is the Bureau of
Labor Statistics, especially its periodic data collections such as the Current Population Survey. Many of these data are directly accessible at the
Bureau of Labor Statistics website (www.bls.gov/oshhome.htm), but
many of the tables and figures are the result of a specific request and can
be replicated only by contacting the Bureau of Labor Statistics and asking for a tabulation of the specific data in question.

REVIEW OF PAST WORKPLACE CHANGES

The workplace has changed in a number of ways that may be important both to the future workplace and to the drawing of inferences about
how these changes are likely to affect the future training needs of OSH
personnel. Several factors are prominent in the changing workplace. The
rapid growth of the number of jobs and the greater proportion of jobs in
the service sector are important changes. Another important change is the
changing nature of the relationship of the worker to the workplace, in that
this relationship is increasingly less permanent or long term. These
changes mean that delivery of OSH training may need to be more associated with the worker and not necessarily delivered just at the workplace.

The U.S. economy has been very dynamic over the last decade with
respect to the labor market. It has, for instance, continued its pattern of
remarkable job growth by expanding by nearly 18 percent and adding
more than 20 million jobs over the period from 1988 to 1998. Although
noteworthy in itself and the envy of much of the industrial world, this
dynamic growth has been accompanied by significant job market restructuring. The industrial and occupational structures of the U.S. economy
are different in important ways from those of a decade earlier.

Industrial Restructuring

The U.S. economy's expansion of the number of jobs over the last decade has not been shared equally by all industrial sectors. The numbers of jobs in some sectors have declined, whereas other sectors have shown remarkable growth not only in terms of the number of jobs but also in terms of their share of U.S. jobs (Table 4-1).

TABLE 4-1 Employment by Major Industry Division in 1988 and 1998 and Projected Employment for 2010 (numbers of employees, in millions)

Industry Group	1988	1998	2010
Total	117.8	138.5	162.8
Nonfarm wage and salary	104.6	125.0	147.9
Goods producing	25.1	25.3	25.3
Mining	0.7	0.6	0.5
Construction	5.1	6.0	6.8
Manufacturing	19.3	18.7	18.1
Service producing	80.6	100.7	122.6
Transportation, communication, and public utilities	5.5	6.5	7.6
Wholesale trade	6.0	6.8	7.8
Retail trade	17.9	22.5	25.2
Finance, insurance, and real estate	6.6	7.3	8.2
Services	26.0	37.6	51.5
Personnel supply services	1.4	3.2	4.9
Computer and data processing services	0.7	1.6	2.8
Health services	7.1	9.8	11.7
Offices of medical doctors	1.2	1.8	2.8
Offices of dentists	0.5	0.6	0.7
Offices of other health practitioners	0.2	0.5	0.8
Nursing and personal care facilities	1.3	1.8	2.2
Hospitals	3.3	3.9	4.5
Social services	1.6	2.6	3.8
Federal government	3.0	2.7	2.6
State and local governments (including public schools)	14.4	17.2	19.7
Agriculture[a]	3.4	3.6	3.6
Private Household Workers	1.2	1.0	0.8
Nonagricultural self employed[b]	8.7	9.0	10.5

[a]Agriculture includes landscaping firms, which account for the increases in this sector, as the increases for landscaping firms more than offset the declines in farm employment.
[b]This group also includes unpaid family workers.

SOURCES: Historical data are from the Bureau of Labor Statistics Survey of Nonfarm Employment, Hours, and Earnings, annual averages, selected years. Projections are by the Committee to Assess Training Needs for Occupational Safety and Health Personnel in the United States, Institute of Medicine, based on the Bureau of Labor Statistics' projections for 1996 to 2006.

Occupational Restructuring

An important dimension of the restructuring of the U.S. economy over the past decade is the fact that the skills required in the declining and the expanding sectors are not identical, and in some cases they are very different. The restructuring of the industrial sector has therefore caused an equally dramatic change in the occupational structure of the economy.

As noted earlier, the economy added over 20 million jobs from 1988 to 1998, or an 18 percent growth. Any occupational group that expanded at a slower rate (all major occupational groups added jobs) saw its share of overall employment decline. Those occupational groups with the slowest growth and the concomitant sharpest decline in share of overall employment were agriculture, forestry, and fishing occupations and the precision production, craft, and repair occupational group, the latter of which is associated, in many instances, with manufacturing. Slower than average growth was also experienced by operators, fabricators, and laborers and also by clerical workers, groups that have seen technology lower the demand for their skills. On the other hand, very rapid growth was seen among the professional specialty and the executive, administrative, and managerial occupational groups. Marketing and sales and technician occupations also showed faster than average growth and thus saw their share of overall employment increase in the past decade. Workers in these rapidly growing occupational groups tended to be employed in very large numbers in the rapidly growing service sector (Table 4-2).

TABLE 4-2 Employment by Major Occupational Group for 1988 and 1998 and Projected Employment for 2010 (in millions of persons)

Occupational Group	1988	1998	2010
Executive, administrative, and managerial	12.1	14.4	17.5
Professional specialty occupations	14.7	19.7	26.4
Technicians and related support occupations	3.9	4.9	6.1
Marketing and sales occupations	12.1	15.5	18.6
Administrative support occupations, clerical	22.1	24.7	27.0
Service occupations	18.4	22.3	27.8
Agriculture, forestry, fishing, and related occupations	3.6	3.8	3.8
Precision production, craft, and repair occupations	14.1	14.5	15.3
Operators, fabricators, and laborers	14.1	18.3	20.2
Total	117.8	138.5	162.8

SOURCE: Historical data are from the Bureau of Labor Statistics Occupational Employment Statistics Survey, selected years. The projections are by the Committee to Assess Training Needs for Occupational Safety and Health Personnel in the United States, Institute of Medicine, based on the Bureau of Labor Statistics' projections for 1996 to 2006.

Given the dramatic shifts in the age, gender, racial, and ethnic compositions of the workforce discussed in the previous chapter, one might expect that this would result in prominent shifts in who holds what type of job. Although changes have taken place and in some instances this compositional change has been dramatic, many industries and occupations still employ either predominantly men or predominantly women. For example, heavy construction and mining still employ primarily men, whereas health technician jobs are mostly held by women. Also, racial and ethnic minority groups are still underrepresented in highly skilled professional and technical jobs. Furthermore, the youngest members of the workforce are most likely to enter the workforce in service industry jobs or in fast-food restaurants and family-owned farms and businesses.

The most significant story in terms of job growth in the past decade, however, has been in the service sector, which added more than 11 million jobs. Several individual industries within this sector have made significant contributions to overall job growth. These include personnel supply services (the temporary help agencies), which added more than 1.8 million jobs from 1988 to 1998. Equally important was the 1.0 million jobs added in the offices of physicians, dentists, and other health care practitioners. At the same time hospitals and nursing homes combined increased their employment by more than 1 million. Also, social service agencies added nearly 1.0 million jobs over this 10-year span. Another significant contributor to job growth was engineering and management consulting services.

A view of structural employment change over the past decade shows that employment share increased significantly only in the service sector. All other major industrial segments either retained about the same share of total employment or had declines in their share of employment (such as for manufacturing). Two important factors have strongly influenced the overall change in the structure of employment in the industrial sector of the U.S. economy. The first of these factors is the intersectoral differences in the growth of productivity. One of the reasons for the decline in employment in such sectors as manufacturing and mining, for example, is the rapid rate of growth in their productivity—particularly when it is compared with those of other industrial sectors. Conversely, the rapid employment growth in many of the service industries has been, at least in part, because of their relatively slower rate of growth in productivity.

The second factor important to intersectoral employment shifts is the shift in the structure of demand. Consumer, business, and government demand for services such as medical, educational, recreational, and computer consulting services is growing more rapidly than the demand for goods such as automobiles, televisions, stereos, and household appliances.

Factors such as globalization of trade are also important (see Chapter 5 for a discussion of this topic).

A LOOK TO THE FUTURE

The committee decided to focus on the period to 2010 to give some perspective on the future but not so far into the future that rapidly changing technology could alter in a radical way the shape of the future trends described here. Still, any look at the future must be done with due respect to possible unforeseen changes that could radically alter trends important to education and training in OSH.

The job growth in the U.S. economy over the past few decades can be expected to continue through 2010, albeit at a somewhat slower pace. This slower rate of increase reflects the demographic slowdown as the newer recruits to the labor force are drawn from a smaller birth cohort than was the case in the previous several decades. Still, the U.S. economy can be expected to add more than 24 million jobs between 1998 and 2010. However, much of the pattern that described the previous decade can be expected to continue. Therefore, by 2010 the mining, manufacturing, agricultural, and federal government sectors are expected to have fewer employees than they did in 1998. Most other sectors, including state and local governments, construction, and retail trade, can be expected to expand employment but at a rate of increase that is unlikely to increase their share of overall employment significantly and that may in a few instances result in a small decline in their share of overall employment. Thus, the service sector, which is projected to add nearly 14 million jobs, is expected to be the dominant player in terms of employment increases. Business services such as computer and data processing and personnel supply firms are expected to be prominent in this job growth picture, adding 1.7 million and 1.2 million jobs, respectively. Health services is expected to grow by 1.8 million jobs, with offices of physicians, offices of other health care practitioners, nursing and personal care facilities, and hospitals each being very important in that growth. Social service agencies are projected to add more than 1 million jobs over the period from 1998 to 2010.

Occupational Growth

Employment in all occupational groups is expected to grow, but it will be at very slow rates for occupational groups associated with sectors in which employment is projected to decline or to grow only moderately. On the other hand, employment in occupational groups associated with sectors that grow rapidly, such as health care, social services, or business services, can be expected to expand greatly. The rate of employment growth in four occupational groups—executive, professional, technician, and service—is projected to grow more rapidly than the overall rate of growth in the U.S. economy (Table 4-2).

Occupational Injuries

Data on occupational injuries and illnesses show two important trends. First, there was a fairly steady decline in the overall rate of occupational injuries over the period from 1984 to 1997.* Second, the rate of occupational injuries is higher in the declining industries such as manufacturing and mining than it is in industries in which employment has grown and is expected to continue to grow (Table 4-3). However, this is not uniformly true. For example, construction is a growing industry, but its rate of occupational injuries, although declining, is still relatively high. Also, although in general the rate of occupational injuries is lower in the service industries, that is not the case for the health services industry, for which the rate is nearly as high as it is for construction or agriculture. Thus, although the changing industrial structure of the economy—all else being equal—should continue to lead to slightly lower overall rates of occupational injuries, there are likely to be some important exceptions. In 1997, truck drivers, those in construction-related occupations, and those in health occupations had among the largest numbers of nonfatal occupational injuries and illnesses (Figure 4-1). Employment in all three of these sectors is projected to grow substantially.

A Census of Fatal Occupational Injuries has been conducted and published by the Bureau of Labor Statistics since 1992 (Table 4-4). Each year between 1992 and 1998 has witnessed more than 6,000 fatal occupational injuries. Although the rate of fatal injuries has declined over this time period, the decline has been a very gradual one and largely one that has been due to an increase in the number of workers rather than a decrease in the number of deaths. More than 40 percent are transportation-related accidents. Assaults and violent acts are responsible for about 16 percent of occupational fatalities, as is contact with objects. Truck drivers, those in

---

*The Survey of Occupational Injuries and Illnesses is a survey of recordable injuries and illnesses conducted by the Bureau of Labor Statistics from data recorded by firms as required by the Occupational Safety and Health Administration. The reported injuries and illnesses are those required by this act to be recorded. Evaluations made by the Bureau of Labor Statistics show both underreporting and overreporting, although underreporting was found to be more prevalent. Overreporting most often occurs with injuries or illnesses not required to be reported. Analyses have not been conducted to determine whether the rate of under- or overreporting has changed over time.

TABLE 4-3 Incidence of Nonfatal Occupational Injuries and Illnesses by Private Industry Division, Selected Years

Industry Division	Total Number of Cases per 100 Full-Time Wage and Salary Workers					Percent Change, 1985–1997
	1985	1990	1995	1996	1997	
Agriculture, forestry, and fishing	11.4	11.6	9.7	8.7	8.4	−23.3
Mining	8.4	8.3	6.2	5.4	5.9	−30.8
Construction	15.2	14.2	10.6	9.9	9.6	−36.8
Manufacturing	10.4	13.2	11.6	10.6	10.3	−7.2
Transportation and public utilities	8.6	9.6	9.1	8.7	8.2	−4.7
Wholesale and retail trade	7.4	9.6	9.1	6.8	6.7	−9.5
Finance, insurance, and real estate	2.0	2.4	2.6	2.4	2.2	10.0
Services	5.4	6.0	6.4	6.0	5.6	3.7
Total private industry	7.9	8.8	8.1	7.4	7.1	−10.1

NOTE: Injuries and illnesses are included in the survey if they are recordable under the Occupational Safety and Health Act's record-keeping requirements.

SOURCE: Bureau of Labor Statistics (1999a).

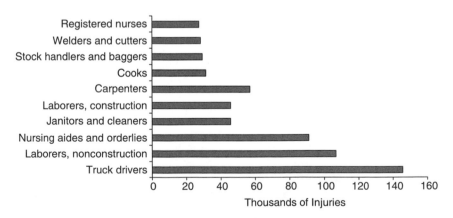

FIGURE 4-1 Occupations with the highest number of nonfatal occupational injuries and illnesses, 1997. Rates are not shown because the data on the number of hours worked by occupation, which are necessary for calculation of rates, are unavailable. SOURCE: Bureau of Labor Statistics (1998a).

TABLE 4-4 Fatal Occupational Injuries, 1992
to 1998, in Private Industry and Government

Year	No.	Rate*
1992	6,217	5.0
1993	6,331	5.0
1994	6,632	5.0
1995	6,275	5.0
1996	6,202	4.8
1997	6,218	4.7
1998	6,026	4.5

*Rate of fatal occupational injuries per 100,000 workers,
civilian, age 16 and older.

SOURCE: Bureau of Labor Statistics (1999b).

farm occupations, those in sales occupations, and construction laborers
have the largest numbers of fatal occupational injuries. Men are more
likely to suffer fatal occupational injuries, as are the self-employed.

Another perspective on injuries can be obtained from the report *Reducing the Burden of Injury* (Institute of Medicine, 1999). That report shows
that injuries (from all sources, not just occupational) have a higher cost,
measured either in dollars or in years of potential life lost, than many
chronic diseases such as cancer, heart disease, or human immunodeficiency virus infection.

Union Membership

The percentage of the workforce that is unionized has been declining
for many decades (Table 4-5). That decline is true not only for the total
workforce but also for all major sectors of the economy except the public
sector. Inasmuch as unions have been an important source of advocacy
for worker protection programs and OSH training, the decline in the
unionized share of the workforce diminishes the reach of this means of
worker training. Unions have recently responded to this decline by increasing their efforts to organize workers, in the health care and public
sectors and among temporary workers in particular.

Contingent and Alternative Employment Arrangements

The workplace has been transformed in other ways that may have
important implications for the training of safety and health personnel in
the future. One of these is the rise of an industry called personnel supply

TABLE 4-5 Percentage of Workforce That Is Unionized by Major Sector, Selected Years, 1983 to 1998

Sector	1983	1985	1990[a]	1995[b]	1998
Total	20.1	18.0	16.0	14.9	13.9
Mining	20.3	17.3	17.9	13.8	12.2
Construction	27.5	22.3	20.6	17.7	17.8
Manufacturing	27.8	24.8	20.5	17.6	15.8
Transportation and public utilities	42.4	37.0	31.5	27.3	25.8
Services	7.7	6.6	5.8	5.7	5.6
Government	36.7	35.8	36.4	37.8	37.5

[a]Adjusted to 1990 population controls.

[b]Beginning in 1994, data are not strictly comparable to those from earlier years because of the 1994 revisions in the Current Population Survey.

SOURCES: Bureau of Labor Statistics, Current Population Survey, annual averages, selected years.

services or, more commonly, temporary help. Other increasingly common workers whose employment situation deviates from the traditional employer-employee relationship are contingent workers, independent contractors, and on-call workers.

Temporary Help Agencies

As can be seen in Figure 4-2, the temporary help (personnel supply service) industry grew from 990,000 employees in 1986 to more than 3.2

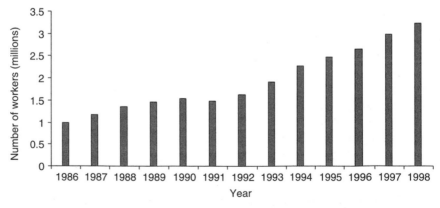

FIGURE 4-2 Employment in personnel supply services, 1986 to 1998 (wage and salary workers in millions). SOURCE: Bureau of Labor Statistics (1998c).

million employed by 1998. This growing industry is one measure of the changing job market, in that many employees are now working for an agency that places them with firms with which there is little expectation of any long-term relationship.

Contingent Workers

The emerging phenomenon of temporary jobs is also exemplified through contingent work. In a survey conducted by the Bureau of the Census for the Bureau of Labor Statistics (Bureau of Labor Statistics, 1995), a contingent worker was defined as one who did not have an implicit or explicit contract for ongoing work. In that survey, three slightly different definitions of contingent work were used, yielding a February 1995 range of estimates of contingent workers from a low of 2.2 percent of all workers to a high of 4.9 percent. The survey revealed that contingent workers were more likely to be young, to be in school, to hold a part-time job, or to be employed in the service industry. Even in the service industry, they represented only a very small proportion of total employment (3.4 to 7.5 percent).

Independent Contractors

The Bureau of Labor Statistics defines independent contractors as wage and salary employees (i.e., not business operators such as shop owners and restaurateurs). The February 1995 survey (Bureau of Labor Statistics, 1995) found that 8.3 million workers (6.7 percent of all employed individuals) said they were independent contractors. Compared with workers in traditional employment arrangements, independent contractors were more likely to be male, white, over 24 years old, out of school, and holders of a college degree. They were somewhat more likely to work part-time and to hold managerial, professional, sales, or precision production jobs. They were more likely to work in construction, agriculture, and services, and were somewhat less likely to be employed in wholesale or retail trade. In contrast to on-call workers and workers employed by temporary help agencies, who generally preferred to be in traditional work arrangements, more than 80 percent of independent contractors preferred their current arrangement.

On-Call Workers

Substitute teachers and construction workers are examples of still another category of employees with alternative employment arrangements. These are people in a pool of workers who are only called to work as needed, although they may be called to work for several days or weeks at a time. Two million workers (1.7 percent of all employed individuals) classified themselves this way in 1995 (Bureau of Labor Statistics, 1995). The demographics of this group, which includes day laborers, were similar to those of workers in traditional employment arrangements, although the on-call group was slightly younger. On-call workers were more likely to be in the services industry and were more than three times as likely to be in the construction industry.

Work at Home

The workplace is changing in another important way, in that for a growing segment of the workforce the workplace is now their home. In addition, many other workers take home work from their regular place of work. A special survey done in May 1997 showed that over 21 million workers reported that they work at home (Bureau of Labor Statistics, 1998d). The survey showed much higher at-home rates of employment for those in white-collar occupations such as professionals and sales workers than for those in blue-collar occupations such as craft workers. As a series of investigations into at-home assembly of electronic components in California's Silicon Valley (Ewell and Ha, 1999) have revealed, however, such arrangements can encourage exploitation of low-paid manual workers by hiding violations of labor and safety laws. Also, the at-home employment rate was higher for those in the service industry, real estate, and wholesale trade. Although this one-time survey does not give trends, every expectation is that the phenomenon of working at home will grow and will further complicate the implementation of OSH programs and enforcement of Occupational Safety and Health Administration (OSHA) standards by putting the workplace in a location that most Americans consider most private.

Number of Jobs in a Lifetime

As the workplace changes in many different ways, one of the common assumptions is that most workers will have many jobs over the course of their working lives. However, it is not possible to track this development by use of data from the sample surveys used for most of the large data collection efforts. It is necessary to follow the same workers through time through the use of what is termed a "longitudinal database." Such a database has tracked a sample of workers from 1978 to 1995, following them from age 18 to age 32.

Although the data in Figure 4-3 do not answer the question of whether this phenomenon is growing, it does show that workers have had a large

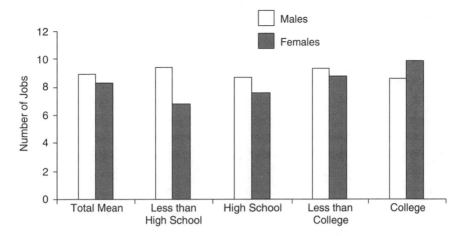

FIGURE 4-3 Mean number of jobs held between the ages of 18 and 32 (1978 to 1995) reported in 1995 by individuals ages 31 to 38 in 1995, by level of education. SOURCE: Bureau of Labor Statistics (1998e).

number of jobs by age 32. However, most of these jobs have taken place by age 22. Women have more jobs with increasing educational levels, whereas the pattern for men by level of education is less clear. The committee's prediction is that these patterns will continue to be characteristic of youth entering the labor market.

Multiple Jobs

The proportion of people who hold multiple jobs has increased gradually from 5.2 to 6.2 percent of job holders over the last two decades. However, the rate of multiple-job holding has gradually been declining among men but has been increasing rapidly among women. In the latest data women represent nearly 44 percent of those who hold multiple jobs, although in 1970 they represented less than 16 percent (Stinson, 1997).

Part-Time Workers

Another workplace change is the share of the workforce that is working in part-time jobs. Although in earlier periods part-time work had grown as a share of employment, in the last few years it seems to have leveled off (Table 4-6). This may in part be related to the fact that in the 1990s the economy was operating at a much higher capacity. In 1998, 25.9 percent of women were employed part-time, whereas 10.6 percent of men were employed part-time.

TABLE 4-6 Employed Persons by Full- or Part-Time Status, 1970 to 1998

| Year | Percentage of All Workers | | |
	Full-Time	Part-Time	Part-Time for Economic Reasons*
1970	84.8	15.2	3.1
1975	83.4	16.6	4.4
1980	83.1	16.9	4.4
1985	82.6	17.4	5.2
1990	83.1	16.9	4.4
1995	81.4	18.6	3.6
1996	81.7	18.3	3.4
1997	82.1	17.9	3.1
1998	82.3	17.7	2.8

*Those who want full-time work but who are unable to find a job. Part-time is defined as <35 hours a week.

SOURCE: Bureau of Labor Statistics, Current Population Survey, annual averages, selected years.

Employment by Size of Employer

An important way of reviewing the dynamics of the U.S. workplace is by an examination of the size of the workplace in terms of number of employees. This measurement can be made in two different ways: at the firm (or total company) level or at the establishment (or individual plant or office site) level. The Small Business Administration maintains a database that is helpful for measuring employment at the firm level. These data show that in 1996, 48 percent of employed people were employed by firms with more than 500 employees. Of the 52 percent who were employed by firms with less than 500 employees, nearly 20 percent were employed by firms with less than 20 employees. More importantly, the dynamics of change are moving employees from the larger to smaller firms. For example, from 1990 to 1995, 1.8 million firms were founded and survived until 1995 (Figure 4-4). However, over the same period, 1.5 million firms that existed in 1990 had failed by 1995, leaving a net addition of more than 250,000 new firms. This added 1.5 million jobs (22 percent of total jobs added over this period), and continuing firms accounted for an additional 5.4 million new jobs. Of these 6.9 million net new jobs created from 1990 to 1995, firms with less than 500 employees provided 76.5 percent of those jobs, and the very small firms (those with less than 20 employees) created nearly one-half of the total (49.5 percent).

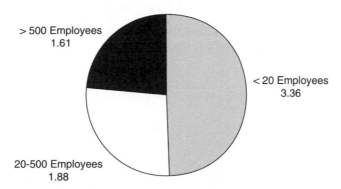

FIGURE 4-4 Aggregate employment increase (in millions) between 1990 and 1995 for firms of various sizes. SOURCE: Small Business Administration, Office of Advocacy (1998).

A review of the net new jobs by sector reveals that of the 6.9 million net new jobs added between 1990 and 1995, 5.8 million were in services, with jobs in health care accounting for more than 2.0 million of that total. In the overall service category, 62 percent of the net new jobs were in firms with less than 500 employees, and 29 percent were in firms with less than 20 employees. However, the split of jobs added in health care services was nearly evenly divided between those firms with less than 500 employees and those with more than 500 employees Additionally, the very small health care services firms provided only 15 percent of the net new jobs from 1990 to 1995.

The Bureau of Labor Statistics also has a database that allows an examination of the numbers of employees at the establishment level. These data are shown in Table 4-7. Large establishments (those with more than 1,000 employees) showed very large decreases in their share of employees between 1980 and 1997. By 1997 less than one in eight workers was employed by an establishment with more than 1,000 workers. This decline has been offset by growth in the share of employees in small establishments, with the largest changes taking place in establishments with 20 to 49 employees and those with 100 to 249 employees. However, with the exception of the very smallest establishments, some increases have taken place in all establishments with less than 250 employees. Some of this shift to smaller establishments reflects the movement of employment from manufacturing to services.

This examination of the dynamics of job creation and the role of small firms in the job creation picture shows that the idea of placing specially trained OSH personnel in firms is probably feasible for only the small

TABLE 4-7 Employment by Size of Establishment in Selected Years, 1980 to 1997

Number of Employees	Percentage of All Private Jobs				
	1980	1985	1990	1995	1997
<5	4.6[a]	6.6	6.3	6.7	6.6
5–9	9.1[b]	7.7	7.8	8.5	8.3
10–19	9.2	9.7	9.9	10.9	10.8
20–49	14.1	14.9	15.1	16.5	16.7
50–99	11.4	12.3	12.3	12.9	12.9
100–249	14.4	15.2	15.8	16.2	16.4
250–499	9.9	9.7	9.8	9.3	9.5
500–999	8.6	7.8	7.9	7.1	7.2
>1,000	18.8	15.8	15.2	11.8	11.7

[a]For 1980 only, this category is less than four employees.
[b]For 1980 only, this category is four to nine employees.

SOURCE: Bureau of Labor Statistics (1998c).

percentage of firms that are very large (and these have declining shares of total employment). Even among establishments, an increasing share of total employment is found in smaller establishments and a declining share is found in establishments with more than 250 employees, with rapidly declining shares found in the largest plants or offices.

IMPLICATIONS FOR OCCUPATIONAL FATALITIES, INJURIES, AND ILLNESSES

Many changes have taken place in the workplace over the last few decades, and many changes are projected in the future. For this report the question is what these changes imply for occupational fatalities, injuries, and illnesses, the implementation of workplace safety and health programs, and for the education and training of OSH personnel.

As seen in Table 4-3, the rate of occupational injuries is high in many industries with declining numbers of employees such as manufacturing and mining. At the same time, in many industries with expanding numbers of employees the injury and illness rates are low. However, there are important exceptions to these broad trends. Construction, an industry with relatively high injury rates, is an industry projected to continue to grow in terms of numbers of workers. Furthermore, the health service industry, unlike many of the other service industries, has an injury rate nearly as high as that of the construction industry or some of the manufacturing industries. This same pattern carries over into specific occupations, where nursing aides have injury rates nearly as high as those of construction workers or some of the production workers in manufacturing and mining. Workers in other occupations that are projected to grow rapidly and that also have particularly high rates of injury or illness include truck drivers, cashiers, and carpenters. Also, the number of upper-extremity strain injuries has been increasing and is particularly high among those in several other occupations expected to grow rapidly such as dental hygienists and meat cutters (Leigh and Miller, 1998). Nevertheless, many other occupations that are found in the service industries and that are also projected to grow have very low rates of injury and illness.

Beyond the changing industrial and occupational structure, many other workplace changes have been discussed in this chapter. More workers are employed by small firms, and that is increasingly true of the new jobs being created. More work is part-time, particularly for women. More workers hold multiple jobs. Some workers now work at home, and a smaller percentage of workers are represented by unions. These changes as a whole describe a significantly changing workplace, one very different from the large fixed-site manufacturing plants with unionized workers in which OSH professionals have made their greatest contributions. Unfortunately, there are not enough comprehensive data to determine if this implies lower or higher rates of occupational injuries and illnesses for the workforce as a whole. However, it is clear that these changes will make it more difficult to identify putative causes of injuries and illnesses across an individual's work history and generally provide OSH services to workers.

IMPLICATIONS FOR TRAINING NEEDS OF
OCCUPATIONAL SAFETY AND HEALTH PERSONNEL

The review in this chapter has noted a number of changes in the industrial and occupational structures of the U.S. economy. Also, the nature of the jobs held by many people has been changing in that the job is more likely to be with a temporary help agency and to be one of multiple jobs for the jobholder. It is also somewhat less likely to be in an industry with a high rate of occupational injury or illness. Furthermore, the dynamics of employment change show that great movement of jobs takes place each year as firms are founded, survive and grow, or fail. This leads to a great deal of movement of workers from one workplace to another. Many of these changes complicate the implementation of workplace safety and health programs, raising the questions of how and where OSH services are best delivered. It may even be necessary to focus delivery of

OSH training directly on the workers. In any case, it argues for types of training and, in particular, delivery systems different from those that have been relied upon to date. For example, a knowledge of and a willingness to work with mass media may be required to reach workers at home as well as at work.

It has always been the case that a large fraction of the U.S. workforce has been outside the sphere of influence of OSH professionals, particularly those professionals who focus on injury and illness prevention. This has principally been because OSH services are most often employed in midsize to large industries and workplaces, but it is also because, in some sectors of the economy, both the workplace and the workforce are transient. If the occupational etiology of workplace illness or injury for most of these workers is recognized, it typically has few consequences beyond the immediate issue of who pays for the medical services that the workers may receive. Neither the elements of clinical case management that relate to an expeditious return to work nor the prevention programs that relate to hazard management and the prevention of reinjury or injury to others exposed to similar risks can be deployed. That is, these workers simply become part of the larger public health problem without reference to or benefit from an understanding of the occupational character of their condition. Some recent efforts by OSHA to reach out to workers like these and to small businesses in general are discussed in Chapter 7.

As has been documented in this chapter, there is considerable evidence that the number of workers outside the sphere of influence of OSH professionals is growing significantly. Simply increasing the numbers or modifying the training of OSH professionals will not be sufficient to address the issue since there is no institutional infrastructure that allows these professionals access to either underserved workers or underserved workplaces. The two principal strategies used to deal more generally with other environmental health problems that are not susceptible to site- or group-specific interventions have been government regulation and public education.

The first of these strategies, government regulation, is represented by OSHA standards. As noted in Chapter 1, there are well over 100 OSHA standards that govern workplace health and safety. There are only 2,488 federal and state OSHA inspectors responsible for enforcing the law at nearly 7 million workplaces (Occupational Safety and Health Administration, 2000). In fiscal year 1999, the 1,242 federal OSHA inspectors conducted 34,342 inspections and the state OSH inspectors combined conducted 54,989 inspections. Two-thirds of these inspections involved construction or manufacturing sites, and the average penalty for the 280,000 violations uncovered was less than $500. These figures represent a decrease of 23 percent in federal inspections and 7 percent in state inspections since fiscal year 1994. The dramatic increase in the $382 million OSHA budget that would be necessary to make inspection a credible deterrent at all 7 million work-sites does not appear likely.

OSHA is nevertheless currently soliciting comments on a draft proposed safety and health program rule that would require all employers except those in construction and agriculture to set up a safety and health program appropriate to the hazards to which their employees are exposed and the number of employees exposed. A model widely used in Europe and Asia goes even further by requiring industry to employ health and safety professionals on the basis of the size and nature of the company. One could take this approach so far as to have the government employ the health and safety professionals and assign them to various industries, as is the Scandinavian practice. Adoption of either of these European models seems highly unlikely given the substantial philosophical, social, and political differences between Europe and the United States. Even the adoption of OSHA's proposed safety and health program rule is far from assured. Given realistic resource forecasts and the contentious nature of regulatory innovation, it is not likely that a traditional regulatory approach will, by itself, succeed in producing a major increase in demand for OSH programs in the heretofore underserved sectors of the U.S. economy. OSHA, in fact, is increasingly focusing on "outreach" programs that provide consultation and training to small business. Some of those efforts are described in Chapter 7.

A more modest and more feasible goal might focus on modification of the many existing standards that include mandates for worker training. Few of these standards say anything about quantity, quality, or efficacy, and as a result, they have had only limited success ensuring effective health and safety training in small workplaces and of the transient workforce. An exception may be that demand for OSHA-mandated worker education, exemplified by the 10-hour general industry and construction safety training courses, has grown dramatically as contractors and owners have included such requirements for hiring of workers or receiving a contract. There is also some precedent for requiring training of managers in federal safety standards. Environmental Protection Agency lead and asbestos abatement training standards, for example, require that supervisors receive the same training as workers, plus some additional training commensurate with their additional responsibilities. A large-scale demonstration project that focuses on high-quality training and evaluation of effectiveness could provide a model and an impetus for expanded worker training programs to meet the needs of the growing workforce underserved by OSH professionals.

The second type of strategy that has been effective in public health, and which is by no means incompatible with broader or more strictly enforced regulation, has used broadly focused educational campaigns carried out through print and electronic media. Examples include promotion of automobile seat belt use, discouraging driving after drinking, promoting healthier eating habits, and decreasing tobacco consumption. To some extent, this strategy is difficult to translate to OSH because it depends for its success on very simple and very general messages; for example, if you use your seat-belt you are less likely to die in a collision, drinking and driving greatly increase your risk of injury or death, or eating lots of fruits and vegetables will decrease your cancer risk. Occupational injury and disease, however, are extraordinarily diverse in their etiologies, and therefore the nature of effective interventions is also diverse, limiting the effectiveness of educational approaches carried out through mass media. Nevertheless, because of the limited mechanisms for reaching those workers who cannot be accessed through midsize to large employers, OSH campaigns in the media should be given renewed attention, perhaps supplemented by the Internet as a means of dealing with the diversity of the messages to be delivered.

If regulation-driven worker education and more focused public education campaigns are deemed a priority in the future as a response to the changes in the workplace that have been outlined in this chapter, increasing demand for specialists in OSH will follow not only in the private sector but also in the public sector, union-based organizations, and academia.

5

The Changing Organization of Work

ABSTRACT. Globalization, technology, and other work-design factors and organizational design innovations present training needs for occupational safety and health professionals. Increasing reliance on computer technology, distributed work arrangements, increased pace of work, and increased diversity in the work environment create several challenges for occupational safety and health personnel. First, potential new hazards may emerge from the introduction of new technologies and through the performance of work in a more boundaryless or virtual organization. Second, businesses are becoming smaller and flatter (fewer levels of management) and are redefining the content of work and the nature of the employment relationship. They are pressured to compete for talent, innovate, provide exceptional service quality, and bring products and services to market fast at competitive prices. The implications of these business developments for workers include demand for new skills and continuous learning, expanded job scope, accelerated work pace, and changing workplaces. Workers face uncertainty in employment relationships, heightened interaction with both customers and other workers, and more involvement with information and communications technologies. Additionally, societal developments like the increasing numbers of single parents, dual-career households, and aged dependents challenge workers to manage multiple and competing interests in their work and home lives. These factors are major sources of time conflicts and carry the potential for dysfunction and distress in the U.S. workforce and at U.S. workplaces.

The committee concludes that occupational health and safety personnel must be knowledgeable about the effects of these changing structural and contextual work conditions on worker well-being and health. They must be competent in recognizing and accounting for the influences of these work organization factors on physical, cognitive, and behavioral functioning, including stress-related conditions and their link to health, safety, and performance. Finally, occupational safety and health personnel must have a basic competence in prevention and organizational intervention strategies and be able to use work organization experts to address workplace stress and well-being issues.

Organization of work or work organization refers to management systems, supervisory practices, and production and service processes and their influence on the way in which work is performed (Sauter et al., 1999). Many if not all of these work organization factors have been shown to affect an organization's culture (Schneider, 1987; Denison and Mishra, 1995) and attitudes toward training, problem-solving, labor relations, and its safety climate (Zohar, 1980; Hofmann and Stetzer, 1996). No data systems in the United States routinely collect information on work organization factors or the numbers and types of organizations and employees exposed to these factors. Nevertheless, it is clear that the past decade has witnessed major changes in both the organization and the nature of work, and it is anticipated that the magnitude and pace of change will continue in response to global competition, advances in technology, and accelerating accumulation of knowledge (Howard, 1995). This fluidity poses important challenges for occupational health and safety training, skills development, and worker safety. The following developments in work organization are most pronounced in large and midsize firms. However, their impacts are increasingly visible in smaller firms that are linked to larger employers in supply chains and service arrangements.

GLOBALIZATION OF TRADE

The accelerated evolution of global trade in the second half of the 20th century has created significant transformations in work organization. Developments in international trade have expanded access to markets, changed the landscape of competition, and increased the mobility of materials, production processes, and investment capital. The creation of a global economy has been facilitated by accords between nations on trade. Of these, the General Agreement on Tariffs and Trade (GATT) and the North American Free Trade Agreement (NAFTA) have been significant for the United States on a global and a regional basis, respectively. The regulatory regimes of GATT have reduced both traditional barriers to trade (e.g., tariffs and import quotas) and nontraditional barriers to trade (e.g., investment policy, local content requirements, and subsidies to domestic producers). More recently, GATT accords have focused on reducing constraints in the flow of both services and goods (Atkinson, 1998).

NAFTA targets specific sectors for tariff reduction (e.g., automobiles, agricultural products, and energy) between the United States, Canada, and Mexico. It prohibits preferential treatment of local companies over ones based in foreign countries, and it limits member governments from influencing any investment or financial service related to trade. Supplemental accords related to NAFTA address environmental concerns, labor, and the subject of import surges (North American Free Trade Agreement, 1994).

Liberalization of trade through global and regional agreements has created a broader context for business decision making. It provides global businesses the opportunity to optimize business performance across the entire enterprise with fewer country-specific restrictions. It does this by enabling greater mobility of plants, equipment, supplies, subassemblies, finished goods, and investment. The impact of enhanced global trade on work organization can be substantial, ranging from wholesale translocation of processes or plants (e.g., *maquilas* in Mexico, Central America, and the Caribbean Basin) to competition-driven changes in performance and efficiency (e.g., automation, changed work processes, new skills, new employment arrangements, and new compensation systems).

Implications of Globalization for Training Needs of Occupational Safety and Health Personnel

Globalization has implications for the training of occupational health and safety personnel. Frumpkin (1998) presented seven issues. First, globalization and free-trade agreements may result in the relocation or displacement of employees by industries or firms. "Unemployment, fear of unemployment, migration, and the accompanying stress and social disruption have a profound impact on the health of workers and their families" (Frumpkin, 1998, p. 237). Second, countries differ in their health and safety standards, and third, they may differ in the degree to which they enforce health and safety regulations. Fourth, hazard communication requires multilingual information and training materials. The fifth issue that Frumpkin raised is the need for trained occupational health and safety personnel. He suggested that less affluent countries have a shortage of trained health and safety professionals. Therefore, organizations with multinational locations may need to provide trained professionals at least to train local personnel. The last two issues dealt with the need to have standardized surveillance data across country borders and the need to disseminate preventive practices and technologies, although not without attention to cultural differences among countries.

WORK DESIGN

The basic arrangement by which work is accomplished has been changing rapidly as a result of global competition and technology. Incorporation of information technology into the workplace, distributed work arrangements, increased hours and pace of work, and diverse cultures in the workforce have all contributed to these changes

Information Technology in the Workplace

Information technology is transforming work along nearly every dimension (e.g., what, where, how, and by whom). Its speed of adoption has been faster than that of other major 20th century innovations. For example, the time that it has taken 30 percent of the U.S. population to become Internet users has been 7 years, whereas it was 17 years for television (U.S. Department of Labor, 1999a). In the United States, the proportion of homes with personal computers grew from 25 percent of all households in 1992 to 50 percent in 1998, and the proportion of all businesses with personal computers grew from 21 to 34 percent during the same 6-year period (John Fisher, International Business Machines Corporation, personal communication, September 3, 1999).

Accelerated use of information technology is enabling business to respond to opportunities and competition from a global marketplace. Figure 5-1 shows that U.S. business investments in computer hardware equipment, for example, have more than doubled since 1990, increasing from $38.9 billion to $95.1 billion in 1998 (U.S. Department of Commerce, 1999). U.S. software sales during the same period increased by a factor of almost three, growing from $50.8 billion to $140.9 billion in 1998 (Business Software Alliance, 1999). In addition, the complexity of information management needs in industry has created an information technology services business for outsourcing, consulting, systems integration, and product support with estimated revenues of $246 billion (John Fisher, International Business Machines Corporation, personal communication, September 3, 1999).

Commerce and work are undergoing major transformations related to this deployment of information technology. The changes are fundamental in nature and are redefining markets, business models, competitors, work methodologies, the concept of workplace, work values, and the life cycle of skills. For example, Internet-enabled "e-business" or "e-commerce" has affected marketing by allowing direct online customer ordering and customization. It has also reduced barriers to entry for new competitors and created competitors of formerly unrelated businesses (e.g., so-called aggregators, which source, bundle, and deliver products

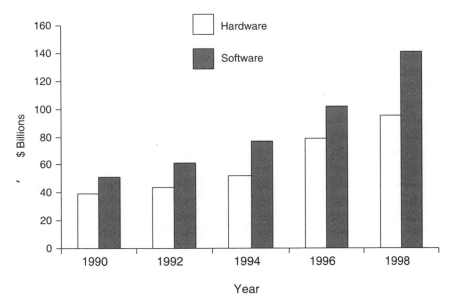

FIGURE 5-1 U.S. investment in computer equipment and software, 1990 to 1998. SOURCES: Business Software Alliance (1999); U.S. Department of Commerce (1998).

produced by others). All industry sectors have been affected by the deployment of information technology because the processes of production, sales, and service are information intensive. Examination of computing technologies in the manufacturing and service sectors illustrates how business, work, and the workplace are being transformed by these technologies.

All core processes in manufacturing have been affected by information technology applications (Figure 5-2). Product design, for example, relies on data collected electronically from sources within and external to the enterprise to define customer requirements. It uses computer-aided designs and simulations to evaluate materials, parts, and assemblies, to test prototypes for performance, reliability, and safety, and to plan for meeting product end-of-life needs such as recycling or reuse. Database integration and analytical software permit assessment of design decisions on quality, cost, and maintenance requirements (National Research Council, 1995, 1999).

Production on the factory floor has expanded the use of computing technology beyond the electronic operation of equipment, sensors, and control systems (robots). Production processes are using information technology to perform real-time tracking of source materials, manage receiv-

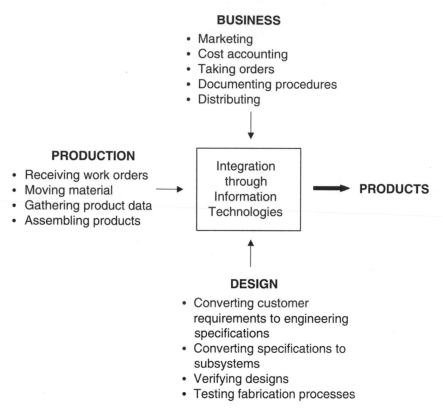

FIGURE 5-2 Information technology as a means of integrating various basic manufacturing activities. SOURCE: National Research Council (1995).

ing and parts flows, schedule production, perform quality control, manage inventories, and speed delivery to customers. Systems applications and networked computing provide an enhanced enterprisewide level of integration in manufacturing involving customers, suppliers, distributors, and other partners. This information-enabled integration has resulted in, for example, integrated product and process design, maximization of work in progress by complementing low inventories and a small labor force with high levels of equipment utilization, and transnational production facilities with networked operations (National Research Council, 1999).

The workforce and organizational effects of these transformations in manufacturing are diverse. At the enterprise level the impact of global competition is creating new business models that promote speed to market and increased flexibility to meet marketplace demands for customized products. These models create businesses composed of a core set of critical competencies that are closely linked to customers and that exhibit high levels of information-driven integration with suppliers, vendors, and other partners. Their link to other aspects of manufacturing, for example, product and process design or production, is varied and dynamic. Outcomes of such business arrangements have included joint venture formations, acquisitions and divestitures, workforce reductions (e.g., downsizing), and outsourcing of support services and production staffs. Competition and technology-enabled customer demands pressure manufacturers to make rapid shifts in their product lines, production processes, and organizations. This promotes changes in suppliers, partners, and mixes of skills. It has also generated a demand for a different type of worker in manufacturing. Employers seek workers who possess learning skills that allow the workers to respond to the shorter half-lives of manufacturing processes, greater cognitive skills so that they can analyze and act on data presented through computer interfaces, effective interactive skills which allow them to work in team environments, and leadership skills so that workers can promote collaboration in cross-functional and multiemployer environments (National Research Council, 1999).

Information technology has also had profound effects on service industries, resulting in improved operational performance and reduced costs. It has been used effectively to simplify and speed the management of large volumes of transactions, automate repetitive tasks, and improve information-intensive logistics activities such as purchasing, inventorying, and distribution. Increasingly, information technology is being used strategically in service industries to create greater differentiation from competitors, enter new markets, or generate entirely new businesses. These efforts include improved information management to customize services, create new or improved products such as online or automated services, and aggregate complementary services for turnkey or "one-stop" shopping (National Research Council, 1994).

Information technology has created an unprecedented ability to uncouple, redesign, and reconfigure service activities. It allows more critical evaluation of each step in a process, each process, the relationships between processes, and the roles and functions of all parties. It has provided business the ability to initiate, eliminate, change, or transfer many service activities, resulting in work restructuring and job redefinition. As a consequence, work in the service industry has both greater routinization and greater standardization, as well as increased autonomy and cognitive (intellectual) complexity. For example, call centers in insurance, banking, and telemarketing have routinized work, whereas mass-customized, short-cycle-time products have increased the knowledge requirements for certain service workers (National Research Council, 1999). Other information technology-related changes in services have involved office automation, flatter organizations (that is, with fewer layers of supervision and management) as a result of information dispersion and collaborative computing, mergers to leverage benefits from technological investments, and outsourcing of information management services. These have contributed to temporary job displacements for many workers, permanent job loss for some workers, and expanded skills requirements for most workers (U.S. Department of Labor, 1999a).

Distributed Work Arrangements

Technology-driven compression of space and time has caused major shifts in concepts of workplace, work space, work hours, and the boundary between work and home. Improved telecommunications, computer connectivity, and computer networks allow work to be done remotely (i.e., at home, on the road, or in client environments) and relatively more independently of the time of day. Although no standard, comprehensive data system that counts workers in nontraditional work environments exists, work at home and on the road (mobile work) is increasing and involves millions of U.S. workers. The Bureau of Labor Statistics reports, for example, that the number of people who work at home for some portion of the workday was in excess of 21 million in 1997 (Bureau of Labor Statistics, 1998d). The number of portable personal computers that have been sold indirectly reflects the enhanced mobile capability of work. Portable personal computer shipments grew from approximately 1.3 million in 1991 to 6.4 million in 1998 (Integrated Data Corporation, 1998). These changes in the location of the workplace create uncertainty regarding potential safety and health hazards, responsibility for their assessment and remediation, the validity of occupational injury and illness surveillance data systems that assume a single employer facility, and traditional notions of "exposure" that are based partly on assumptions of people fixed in space and time.

Changes in information technology hardware and software have altered the tools that people use and have contributed to changes in the physical space in which they work. Fixed computer workstations, mobile computers, other electronic interface devices (e.g., personal digital assistants), and software pose cognitive, biomechanical, human factors, and work organization challenges for comfort, well-being, and productivity. Poor workstation arrangements or setups, excessive work flows and pace, and other factors have been linked to the increased incidence of work-related musculoskeletal disorders (Bernard, 1997). The miniaturization of hardware and the evolution of distributed computing have supported changes in workspaces. Open-landscape designs are replacing enclosed offices, spacesharing among mobile workers and telecommuters is increasing, and more work is being done in client environments, at home, and in travel accommodations. Advances in information technology have also created the potential for time spent working to increase (e.g., at home), for the boundary between home life and work life to blur, and for worker performance to be measured electronically or monitored. A survey by the American Management Association (1999) found that 67 percent of major U.S. firms engage in some form of electronic oversight of employee activity. Conversely, by making work portable, information technology has also provided the opportunity for some workers to achieve a more satisfactory balance between home life and work time.

Caudron (1992) reported that from 1988 to 1992, the number of telecommuters in the United States rose from 3 million to more than 6 million people. Many but not all of these people are working in their homes. These distributed or networked employment relationships are exemplified by other situations, such as engineers who work on assignment as part of a client team, professionals who are members of advisory boards or consortia as representatives of their employers, and clerical workers who work for a number of client organizations over a year. Even within more traditional work sites, individuals are increasingly moving from one workstation to another. The growth in the number of contingent workers and other networked employees makes work environments more fluid and makes the work being performed more fluid.

Hours and Pace of Work

Since 1992, when Juliet Schor published her book entitled *The Overworked American*, it has generally been accepted that people are working longer hours. However, this conclusion has met with debate, at least in the academic community (U.S. Department of Labor, 1999b). Citing the Current Population Surveys of the Bureau of Labor Statistics, the *Report on the American Workforce* (U.S. Department of Labor, 1999b) found the average number of hours worked per week to be relatively stable from 1960 to 1998, varying between 38 and 40 hours per week. However, there were some differences depending on certain subpopulations. For example, with the increased participation of women in the workforce, it is not surprising to find that there has been a small upward trend in the average number of hours worked per week. Although the average number of hours worked per week has been relatively constant for men, there was a small increase in the proportion of men working more than 40 hours per week. Lastly, the number of couples in which both individuals are working long hours has increased (U.S. Department of Labor, 1999a).

Data on shift work collected in a supplement to the May 1997 Current Population Survey (Bureau of Labor Statistics, 1998f) revealed that about

25 million full-time wage and salary workers had flexible work schedules that allowed them to vary the time they started and ended work. That constituted 27.6 percent of workers, up sharply from the 1991 figure of 15.1 percent. Flexible schedules were more likely to be reported by men, parents, sales workers, executives, supervisors, and managers and by workers in service-producing rather than goods-producing industries. The same report (Bureau of Labor Statistics, 1998f) revealed that 15.2 million persons (16.8 percent of all full-time wage and salary workers) normally worked a shift other than one on a regular daytime schedule. Alternate schedules included evening shifts (4.6 percent), night shifts (3.5 percent), rotating shifts (2.9 percent), and employer-arranged irregular schedules (3.9 percent). Such schedules have long been known to be associated with increased stress and disease prevalence and may be associated with an increased frequency or severity of workplace injury as well (Office of Technology Assessment, 1991). The rapid development of e-commerce in the late 1990s and the resulting need for 24-hour responsiveness may increase the number of shift workers, but the 1997 prevalence is only slightly higher than that in 1985 (15.9 percent) and is actually lower than that in 1991 (17.8 percent).

Another time-related aspect of work is the pace of work. With the introduction of computer-aided technology, there has been an increase in the speed with which tasks are completed (Shaiken, 1985). Many of the changes in work procedures, in addition to computer-aided technology, are specifically designed to increase the speed with which organizations and individuals can respond to the demands of production (Wall and Jackson, 1995). This has essentially increased the intensity of the work pace for many employees. Such intensification of work has resulted in increased reports of stress and anxiety (Zikiye and Zikiye, 1992).

Diverse Cultures

Much of the knowledge base concerning the effects of work on employees' safety and health has been based on white, American-born, male workers. As pointed out in Chapter 3, the workforce is becoming much more diverse. This diversity may be within groups working side by side on the shop floor, in professional and technical offices, and in the boardroom. There is some evidence that there are gender and cultural differences in people's responses to technology. For example, Hackett and colleagues (1991) reported that women had more negative expectations than men about changes in their working conditions with the introduction of technology even after controlling for education, seniority, age, and other relevant characteristics of the job. There also may be intercultural differences in reactions to technology (Coovert, 1995). Immigrant workers, collaborative work across national boundaries using Internet technology, and international alliances and mergers are all contributing to work in multicultural environments. This work environment requires accommodations and adjustments to meet a heterogeneous array of individual and group needs to successfully exploit ideas and innovations.

Implications of Changes in Work Design for Training of Occupational Safety and Health Personnel

Technology, distributed work arrangements, increased hours and pace of work, and increased diversity in the work environment create special challenges for occupational safety and health personnel. First, although application of automated production methods to dangerous and repetitive tasks has reduced some traditional workplace hazards, new hazards may emerge related to a specific technology (e.g., emissions or cognitive demands) or the manner in which it is introduced or deployed. Second, distributed work arrangements create a broad array of largely unregulated work environments outside the normal practice of occupational safety and health personnel. Third, enhanced workforce diversity challenges homogenized (i.e., "one size fits all") communications. Understanding a workforce's diversity will be key to developing training programs and implementing interventions that address diversity-related beliefs, customs, or practices that affect health or safety. Last, the emergence of cognitive load, the increased work pace, and the increased work hours create an urgent need for occupational safety and health personnel to be as familiar with the psychosocial aspects of work as they are with the physical, chemical, and biological aspects.

ORGANIZATIONAL DESIGN

The changing nature of work is not occurring in a vacuum. Many of these changes are results of organizations' attempts to be more productive and competitive in a global market. This has prompted changes in organizational structures and management systems.

Self-Managed Work Teams

Manufacturing organizations started experimenting with self-managed work teams in the 1960s. By the 1990s, 26 percent of organizations surveyed by Wellins and colleagues (as cited in Wellins et al., 1991) reported using teams, but 59 percent of the organizations that reported using teams indicated that less than 10 percent of their employees were structured in teams. Despite this, the executives in that study expected that more than half of their employees would be working in self-managed teams by the mid-1990s. One finding that is particularly relevant to the training of occupational safety and health professionals is that in that study 69 percent of the organizations that used the team concept had placed the responsibility for safety with self-managed teams.

More than 70 percent of the organizations that participated in the study of Wellins and colleagues were in manufacturing; however, Wellins and colleagues estimated that teams were being introduced widely in all sectors of the economy. Mohrman and Cohen (1995) indicated that project teams were being used extensively in many industries, including "almost all companies that have a new product development process" (p. 371). Proehl (1996) refers to a 1992 *Training* magazine article that reported that 80 percent of U.S. organizations with 100 or more employees used teams and that in the largest organizations, those with 10,000 or more employees, the figure was closer to 90 percent. Coupling the notion of a matrix organization that integrates lateral communication channels with the hierarchical communications channels of more traditional organizational structures with project teams has resulted in the growth of cross-functional teams.

"Quality" Management Systems

The search by organizations to be more competitive has resulted in several new approaches to managing business processes such as just-in-time (JIT) inventory control, total quality management (TQM), and advanced manufacturing techniques (AMT). Although it is not known how many organizations are using one or more of these approaches, a survey found that almost 70 percent of the manufacturing companies in the United Kingdom were using JIT inventory control techniques as of 1991 and that most of these companies had implemented the technique since 1988 (Oliver and Wilkinson, 1992). Similar patterns of widespread use and recent shifts to these approaches are seen for TQM and AMT.

All three of these approaches are meant to enable organizations to provide a more rapid and better quality response to the demands of production. Furthermore, these processes increase the cognitive demands on employees, the responsibilities of employees, and the interdependence among all employees in the plant. As Wall and Jackson (1995) point out, these changing demands on employees can result in job-related strain. Joyce (1986) also found that changing to a matrix organization increased the level of role ambiguity, a risk factor for job stress. The experience of job stress and strain has been linked to negative effects on people's physical health (e.g., hypertension and cardiovascular disease, [Schnall et al., 1998]), as well as mental health (e.g., depression and anxiety [Quick et al.,

1992]). However, these approaches are espoused to empower employees and increase their job control. If this is the case, then there should not be an increase in job stress and workers' mental health would be improved (Karasek and Theorell, 1990). Little research has directly assessed the effects of these approaches to organizational systems on employees' health. Landsbergis and colleagues (1999), in their review of the limited research, found little support for the health-enhancing increase in empowerment and suggested that the work intensification that appears to be frequent in these systems may actually have a negative effect on the physical and mental well-beings of workers.

Nonstandard Employment Agreements

Nonstandard work arrangements include regular part-time employment, self-employment, temporary help agency work, independent contracting, and on-call work. These work arrangements antedate regular full-time employment in the United States (Jacoby, 1985) and in 1995 were estimated to account for 2.2 to 4.9 percent of total U.S. employment (Bureau of Labor Statistics, 1995). The largest numbers of nonstandard workers perform part-time work and are self-employed, but temporary help agency work has exhibited the greatest rate of growth during the last decade (see Figure 4-2 in Chapter 4). Although nonstandard work arrangements account for a small proportion of overall employment in the United States, the absolute number of workers is large and the proportion of companies that use these forms of labor has increased (Fierman, 1994; Cooper, 1995; Filipczak et al., 1995). Firms use contingent workers for a variety of reasons, including flexibility to manage variations in demand, acquiring specific expertise, filling in for absent employees, and controlling head counts or other expenses because of downsizing or competitive pressures (Rousseau and Wade-Benzoni, 1995; Human Resources Institute, 1996).

Nonstandard work arrangements create a work context with special challenges. Temporary help agency workers and independent contractors work in client environments and shift their work environments at greater frequencies than noncontingent workers (National Research Council, 1999). Control over these environments often resides outside the individual or his or her employer, and changes to the work environment may involve financial, legal, and other complexities not present in fixed, single employer premises. Occupational safety and health training is particularly important in this work context, but the person or organizational unit responsible for providing it may not be clear and contingent workers may be at risk for not receiving it in a timely fashion or in adequate amounts. For example, Kochan and colleagues (1992) reported that one-third to one-half of contract workers in the petrochemical industry were excluded from plant safety training programs. Access to health care may be an issue with contingent workers since they tend to have fewer or no health benefits and tend to receive lower wages than full-time workers (Barker, 1995). Growth in the numbers of contingent workers will challenge OSH personnel to rethink notions of exposure, surveillance, and intervention strategies, given the fluidity of the work environment. In addition, understanding legal considerations in multiemployer work environments will be important for occupational health professionals in implementing work site training and providing appropriate recommendations for workplace modification.

Downsizing

Since the 1980s the effects of increased international competition and globalization plus the impacts of technology have prompted organizations to consider downsizing, restructuring, and reengineering as business strategies rather than simply responses to organizational decline (Burke and Nelson, 1998; Martin and Freeman, 1998). Palmer and colleagues (1997) suggested that in the United States the reason for downsizing had moved away from general economic conditions to better staff utilization, outsourcing, plant closure, mergers, automation, and the use of new technology.

Cascio (1993) reported that between 1987 and 1991 more than 85 percent of the Fortune 1000 firms downsized their white-collar employees. These events affected more than 5 million people. According to a Conference Board survey of firms with more than 10,000 employees, 64 percent of the firms that reported that they had downsized experienced a decrease in morale among survivors, 46 percent experienced an increase in retiree health costs, and 30 percent experienced an increase in overtime (Fair, 1998). Furthermore, Pearlstein (1994) reported that two-thirds of the organizations that downsized did so again within a year. Therefore, many have concluded that downsizing, rightsizing, restructuring, or reengineering will continue.

The health and safety effects of downsizing are multiple. Victims of downsizing, those who lose their jobs, frequently experience negative effects on their general physical and psychological well-being (Kozlowski et al., 1993). Fortunately, many of these negative effects of unemployment appear to abate once an individual regains employment. It has also been documented that the survivors of downsizing experience stress and stress-related illnesses such as high blood pressure, dizziness, and stomach upset (Burke and Nelson, 1998). A recent study conducted by de Vries and Balasz (1997) found that several of the individuals they interviewed and who were responsible for implementing downsizing efforts exhibited depressive symptoms themselves, much the same as has been observed among victims and survivors. These results are consistent with a 1996 American Management Association Survey that found that job elimination was associated with increased disability claims. The increase in claims was greatest for categories reflective of stress (e.g., mental or psychiatric problems and substance abuse as well as hypertension and cardiovascular disease).

Implications of Organizational Design for Occupational Safety and Health Personnel

Changes in organizational structure and systems can affect employees' well-being. These effects may be the result of job insecurity, lack of training, or stress and other psychological factors. Occupational safety and health personnel need to be aware of organizational and work design constructs and how they may be affecting employees' health and safety. In addition, occupational safety and health personnel themselves are not immune to job elimination or being assigned to self-managed work teams. Therefore, they need to be skilled in working within multidisciplinary teams and have a core competency in business process, finance, planning, and management.

WORK-LIFE BALANCE

Social, economic, and demographic changes in the United States have increased the difficulty and complexity of balancing work demands or aspirations and home life. The aging population and increased life expectancy have contributed to the prominence of caregiving among employed persons. The U.S. Department of Labor has reported that two of three caregivers, involving 5.6 million households, were employed in 1996 (U.S. Department of Labor, 1999a). Furthermore, it is estimated that elder care will involve 42 percent of all workers by 2002 (Galinsky and Bond, 1998). Parenting is challenged by the economic necessity for dual-income households (the proportion of households in which both members of the couple work rose from 39 to 69 percent between 1970 and 1998) (U.S. Department of Labor, 1999a). Time demands for women have been exacerbated by the rise in the number of single-parent families since 1970 (11 to 27 percent), the significant increase in the proportion of mothers in the workforce (47 percent in 1975 versus 72 percent in 1998) (U.S. Department of Labor, 1999a), and the rise in the number of women holding multiple jobs (Amirault, 1997; Stinson, 1997).

Progressive employers and those in competition for particular skills have responded to these work-life issues with an array of programs, resources, and accommodations. These include, for example, elder and child care initiatives, flexible work schedules, job sharing, work-at-home arrangements, on-site or telephonic concierge services, and time and stress management training. Most U.S. workers, however, do not have access to such programming or work benefits. The 1993 Family Medical Leave Act's right to job-protected time off from work may be the principal source of temporary accommodation for many workers. However, it is unpaid leave and pertains only to serious medical conditions that affect the employee, a spouse, a dependent, or a parent and to care for a child after birth or placement for adoption or foster care.

IMPLICATIONS FOR OCCUPATIONAL SAFETY AND HEALTH EDUCATION AND TRAINING

The changes occurring in the structure of organizations and the context of work are creating new challenges for workers and worker well-being, particularly psychological well-being. Organizations are becoming smaller and flatter and are redefining the content of work and the nature of the employment relationship. They are pressured to compete for talent, innovate, provide exceptional service quality, and bring products and services to market fast at competitive prices. The implications of these business developments for workers include demand for new skills and continuous learning, expanded job scope, accelerated work pace, changing workplaces, uncertainty in employment relationships, heightened interactivity in job performance, and greater interface with information and communications technologies.

Work stress and its effect on health are major consequences of the changing work environment and the organization of work. Occupational health professionals must be competent in recognizing psychological conditions related to stress and in knowing the appropriate interventions and when to use relevant expert resources. They must also be familiar with concepts that relate to the structure and context of work and organizations to recognize and account for the influences that work and organizational context have on workplace injury, illness, or other health issues. These skills are essential to respond to various manifestations of cognitive overload or other work or organizational sources of distress in the workplace. The need for competency in this area is particularly keen when traditional risk factors fail to explain the occurrence or distribution of cases or events. It has been demonstrated, for example, that work organization factors are important in the epidemiology of work-related musculoskeletal disorders (Moon and Sauter, 1996; Bernard, 1997). Core curricula and training for OSH health professionals need to address these important competencies.

6

The Changing Delivery of Health Care

ABSTRACT. Treatment of workplace injuries and illnesses takes place within the larger U.S. health care system, which has been undergoing dramatic changes during the past decade. One of the most striking features of health care reform has been the dramatic growth of managed care, a major element of which is tighter control on the utilization of health care services. This has led to emphases on seeing more patients (relative to preventive services, education, and research), the use of primary care physicians and other health care professionals instead of specialty-trained physicians, and the consolidation of health care into large occupational health clinics and integrated systems that provide full-time (24 hours a day, 7 days a week) coverage of workers and their families. The committee concludes that all physicians need to be more familiar with workers' compensation and that aspiring occupational health professionals need training in managed care principles and multidisciplinary health care.

Changes in the delivery and financing of health care can have significant effects on occupational safety and health and on those who work in the field. Although some occupational safety and health workers have little contact with the health care delivery system (e.g., safety engineers), others (e.g., physicians and nurses specializing in occupational health) work in that system on a regular basis. More importantly, injured workers receive their care from health professionals and institutions that are mainstream elements of the U.S. health care system: emergency rooms, urgent-care facilities, ambulatory care and occupational health clinics, hospitals, group practices, and individual practitioners' offices. The types of physicians involved in workers' compensation cases can reflect the entire spectrum available, including but not limited to all surgical specialties (orthopedists, neurosurgeons, hand surgeons, burn surgeons, etc.), emergency care physicians, family practice physicians, radiologists, physiatrists, internists, and so forth. Nurses and nurse practitioners are frequently involved in the treatment of work-related injuries, especially the minor ones. Finally, other practitioners are also involved in providing care to injured workers: chiropractors, physical therapists, occupational therapists, and acupuncturists, to name only a few. One of the continuing issues between injured workers and employers has been the control of who selects the health care providers.

Changes in the ways in which health care is delivered thus have the potential to affect not only workers' access to care but also the cost and quality of that care. The U.S. health care system underwent dramatic changes during the decade of the 1990s. These changes were fundamental in nature and had enormous effects. They occurred rapidly and without an overall "game plan" to guide them. Three factors catalyzed the changes: the rising cost of health care, a lack of access to health care by a large and growing segment of the population, and concerns over the quality of health care.

In regard to cost, the U.S. health care system is the most expensive in the world, outstripping by more than half again the health care expenditures of any other country (Iglehart, 1999a). In 1997, U.S. health care expenditures totaled $1.092 trillion, representing 13.5 percent of the gross domestic product (GDP) (Levit et al., 1998) (Figure 6-1). Of the total, 46 percent was spent by federal, state, and local governments, with the private sector financing the rest. Premiums paid by employers and employees to purchase health insurance constituted 60 percent of the component paid by the private sector (Levit et al., 1998). In contrast to the 1980s and the early 1990s, when the annual rate of increase in U.S. health care expenditures was often in double digits, there has been a decrease in the rate of growth during the past 5 years. In 1997, for example, U.S. health care spending rose only 4.8 percent—the slowest rate of growth in more than 35 years (Iglehart, 1999a). Many are predicting, however, that this slower rate of growth will not continue. A recent study by the Health Care Financing Administration has concluded that national spending on health care will again increase at a faster rate than the rate of growth of the economy as a whole and that by 2002 it will total $2.1 trillion, or 16.6 percent of GDP (Smith et al., 1998), perhaps creating additional pressure to cut costs.

Even though the United States spends almost one-seventh of its GDP on health care, a rising number of Americans either lack health insurance or are underinsured. In 1998, 44.2 million Americans had no health insurance. This figure represented 16.3 percent of the population, an increase

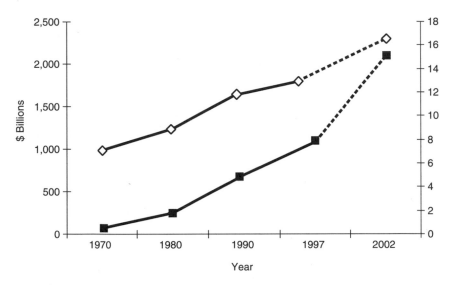

FIGURE 6-1 National expenditures for health services and supplies, 1970 to 1997 and projections for 2002, in billions of dollars (■) and as a percentage of GDP (◊). SOURCE: Levit et al. (1998).

from 14.2 percent in 1995 (U.S. Census Bureau, 1998). Lack of insurance is closely correlated with low income (Kuttner, 1999a). Many Americans are also underinsured and must either forgo care or pay for it out of pocket. A number of recent studies have documented that the percentage of individuals who lack insurance is rising (Kuttner, 1999a).

Increases in the numbers of uninsured and underinsured individuals appear to be more closely related to a deterioration in employer-based coverage than to unemployment, since they have occurred at a time when the latter is declining (Schoen et al., 1998). Two-thirds of Americans receive their health insurance through an employer (Fronstin, 1998). The rising cost of health care has forced many employers to either drop health insurance coverage for their employees or eliminate some of the benefits that they provide. In addition, trends toward part-time or temporary employment, the growth in the number of self-employed individuals, and the contracting out of certain tasks have left many workers and their families with no or inadequate coverage. Other factors include the loss of Medicaid coverage because of welfare reform and the rising cost of "Medigap" coverage, which results in inadequate insurance coverage for elderly individuals enrolled in the Medicare program (Kuttner, 1999a).

Concerns about the quality of U.S. health care have also been a factor in promoting health care reform (Chassin et al., 1998; Bodenheimer, 1999).

Quality has been defined as "the degree to which health services for individuals and populations increase the likelihood of desired health outcomes and are consistent with current professional knowledge" (Lohr, 1990, p. 21). Poor-quality health care can occur as a result of the overuse, underuse, or misuse of health care. Although many believe that the quality of medical care for acute conditions in the United States is the best in the world, there are also substantial data to indicate that each of the factors mentioned above is still a significant problem, resulting in unnecessary morbidity and mortality (Bodenheimer, 1999). In recent years there has been a growing concern that attempts by managed care plans to reduce costs have led to underuse. There is also concern that the U.S. health care system has placed too much emphasis on the care of illness or injury after it occurs and too little emphasis on disease and injury prevention and the promotion of healthy lifestyles (Fielding, 1994).

The failure of either state governments or the federal government to enact or implement comprehensive health care reform legislation has left the private sector in control of many of the changes that are taking place in the U.S. health care system. These marketplace-initiated changes have occurred rapidly and, lacking a national consensus on what the health care system of the future should look like, have occurred in a relatively decentralized and uncoordinated fashion. They have also been controversial (Ginzberg, 1995; Angell and Kassirer, 1996; Schiff, 1996; Schroeder, 1996; Altman and Shactman, 1997; Bodenheimer and Sullivan, 1998; Blumenthal, 1999). Most observers credit them with controlling the rapid rise in health care costs but believe that they have done little to improve access to health care. There is considerable disagreement regarding their impact on the quality of health care.

ELEMENTS OF HEALTH CARE REFORM

One of the most striking features of health care reform has been the dramatic growth of managed care. There are several types of managed care organizations, including health maintenance organizations (HMOs), preferred provider organizations (PPOs), point-of-service plans (POSs), and combinations of the three. The simplest form of an HMO is a health plan that generally has its own health care clinics and providers, who are paid a salary to treat plan members. The HMO members in turn pay a fixed premium for membership and incur few if any additional costs for services received. PPOs are health plans that contract with a limited number of health care providers to provide services to plan members. Providers offer services to members on a fee-for-services basis, usually at a discount from the fees that prevail in the community. Both HMOs and PPOs typically require members to see a designated primary care provider or other generalist before seeking the services of a specialist. POS health plans are variations of PPOs that characteristically allow a member far more freedom to choose providers, but copayments and deductibles are higher for services rendered by providers not in the plan's network of preferred providers. Numerous other variations on these basic models exist, but common to all types of managed care organizations is tight control on the utilization of health care services. Fuchs (1997) attributes the growth of managed care organizations to two factors: rising health care costs and overcapacity in the health care delivery system. It has been estimated that 75 percent of Americans with private health insurance are enrolled in managed care plans (Iglehart, 1999a). Medicare and Medicaid have also been moving their enrollees into managed care.

Consolidation

Marketplace competition for the delivery of health care services has led to the consolidation of provider organizations. One type of consolidation, known as "vertical integration," combines physicians and other health care professionals with hospitals, rehabilitation units, long-term-care facilities, mental health and substance abuse programs, and health promotion and disease prevention programs into an integrated whole that can deliver a comprehensive array of services. Financing mechanisms are often incorporated into these "one-stop-shopping" organizations. In addition, through mergers and buyouts there has been horizontal consolidation of health plans and integrated delivery systems into larger and larger provider entities. Through vertical integration and horizontal consolidation, provider organizations obtain marketing advantages because of a larger area and scope of coverage, have greater access to capital, and obtain economies of scale in things such as purchasing and the development of information systems. Occupational health has not escaped the consolidation movement, and recent years have witnessed an increasing concentration of occupational health and medical services in hospitals and large clinics, often part of regional or national networks.

For-Profit Health Care Organizations

Associated with consolidation in the health care industry has been the growth of for-profit health care delivery organizations (Kuttner, 1999b). Some of these investor-owned entities are newly formed, whereas others represent the conversion of not-for-profit entities to for-profit status. In 1997, 62 percent of HMO members were enrolled in investor-owned health plans (Interstudy, 1998). In the early 1990s there was rapid growth in the number of for-profit hospitals, but this trend has plateaued in recent years. As of 1997, 16 percent of all U.S. hospital beds were investor owned (Kuttner, 1999b). For-profit status has been useful to the extent that it facilitates access to the capital markets, but it has been controversial because it raises the question of conflict between investors' desire for a return on their investment and the provision of quality health care (Kuttner, 1999b).

Reimbursement

There have also been important changes in the methods of reimbursement for the provision of health care. Under the traditional U.S. health care system, providers received fee-for-service reimbursement on the basis of their charges or costs. Under the new system, payment is more likely to be through either capitation or a tightly managed fee for service at "negotiated" discount rates. Capitation places the provider at risk and discourages the provision of health care services by paying only a fixed amount per plan member regardless of the services that the member requires. An additional effect of capitation can be seen in workplaces where treatment for non-work-related injury and illness is covered through capitation and treatment for work-related injuries and illness is covered through a fee-for-service system. Under such circumstances there can be a strong incentive to classify injuries or illness as work related (Ducatman, 1986). A managed fee-for-service system also controls the provision of health services by requiring preauthorization or a second opinion before a service is delivered. A number of significant changes have also been made in Medicare and Medicaid (Iglehart, 1999b,c), including an effort to slow the previously rapid rate of growth of both programs.

Hospitals

In an effort to reduce health care costs and because of the availability of new, less invasive technologies, the site of delivery of most health care has shifted from the hospital to a variety of noninpatient settings such as clinics, physicians' offices, outpatient surgery facilities, and the home. Occupancy rates in U.S. community hospitals fell from 64 percent in 1990 to 60 percent in 1997 (Iglehart, 1999a). Although hospital spending is still the largest single component of U.S. personal health care expenditures, it is also the slowest-growing component of an expenditure survey conducted by the Health Care Financing Administration (Iglehart, 1999a).

Specialists and Primary Care Providers

Reform of the health care delivery system has also affected the providers of health care. In the case of physicians, there has been less emphasis on specialists and specialty care and more emphasis on primary care and the role of the generalist physician (Institute of Medicine, 1996). In some cases, managed care has used generalist physicians as "gatekeepers," without whose approval a patient cannot see a specialist. The number of other health care professionals such as nurse practitioners and physician's assistants has increased rapidly, and their scope of practice has been expanded to include responsibilities that were previously the exclusive domain of the physician (Cooper et al., 1998a,b). This expansion of scope of practice is one example of the phenomenon of "substitution," in which providers with less formal preparation are substituted for those with more education and training in an attempt to reduce costs. Other workforce changes include cross-training and multiskilling of health care workers and the increased use of team delivery of health care (D'Aunno, 1996).

Health Care Practices

The nature of health care has also been changing. Efforts to reduce costs and public interest in improved health have fueled a renewed emphasis on population medicine, health promotion, disease prevention, and greater integration of the various health care disciplines and public health (Rundall, 1994). There is little controversy about the relationship between lifestyle risk factors (e.g., sedentary lifestyle, tobacco use, poor nutrition, obesity, and high lipid levels) and poor outcomes that result in increased morbidity and mortality. The economic impacts of these health risks are substantial (Pelletier, 1996; Goetzel et al., 1998). Furthermore, these risk factors can be modified and workplace health promotion programs can exert a long-term positive influence on health and lifestyle practices (Heaney and Goetzel, 1997). Consequently, the growth in health promotion and disease prevention programs and counseling efforts at the work site have been both substantial and positive (Pelletier, 1996). These efforts are likely to increase as modification of risky health behaviors and cost containment are emphasized (Rogers, 1994; Kosinski, 1998).

Traditional medical practice is becoming more standardized because of evidence-based medicine, clinical practice guidelines, and disease management techniques (Ellrodt et al., 1997), but public interest in natural remedies and in alternative medicine has resulted in significant growth in the number of nontraditional health practitioners and in the widespread use of nontraditional remedies such as herbal medicines (Cooper et al., 1998a,b; Eisenberg et al., 1998; Kaptchuk and Eisenberg, 1998).

Finally, the aging of the U.S. population has increased the emphasis placed on the management of chronic disease.

Academic Health Centers

Finally, changes in the health care delivery system have had a major impact on U.S. academic health centers (AHCs). These institutions educate and train most of the nation's health care professionals, perform large amounts of biomedical research, and provide highly specialized care in their teaching hospitals. AHCs are experiencing difficult times, in part because their traditional emphasis on the education and training of specialists and subspecialists is often out of line with the marketplace, which is emphasizing primary care and generalism. In addition, the single largest source of revenue for medical schools, the major components of most AHCs, is income derived from patient care—either through faculty practice plans or via transfers from teaching hospitals (Jones et al., 1998). It has been shown that AHC hospitals are approximately 30 percent more expensive than their nonteaching counterparts (Reuter and Gaskin, 1997). In a reformed health care system that places great emphasis on cost containment, many AHCs are having a difficult time competing for patients. To date only one AHC has been forced to close, but others are merging their programs to reduce costs and increase efficiency. If AHCs encounter severe financial difficulty in the future, their ability to maintain their programs of teaching, research, and patient care could be jeopardized. One result could be impairment of their programs in occupational safety and health education, training, and research.

Managed Care in Workers' Compensation

Nationwide, the costs of workers' compensation doubled between 1985 and 1990 (Soloman, 1993), and beginning in 1991 states began to pass legislation that allowed or required workers' compensation programs to copy the cost-control techniques of managed care group health plans. The earliest efforts at cost control involved the implementation of discounted fee schedules for providers and health facilities. A 1996 study by the Workers Compensation Research Institute (Burton, 1996) found that fee schedules had been introduced in 40 states and that hospital billing regulations had been introduced in 35 states. From the little research available, it appears that fee schedules have not reliably reduced medical costs, as providers in the predominantly fee-for-service systems delivered more or more complex services per case, offsetting the lower fees for each service imposed by the schedule (Nikolaj and Boon, 1998).

Subsequent approaches to managing health care in workers' compensation cases have turned to networks of preferred providers, treatment guidelines, aggressive case management, and 24-hour coverage (Dembe, et al., 1998; Leone and O'Hara, 1998; Nikolaj and Boon, 1998; Weinper,

1999). Nikolaj and Boon (1998) report that 48 states now limit workers' choice of providers in some way (the employer or the insurer chooses the provider or restricts the employee's initial selection or freedom to change providers). There are as yet no definitive answers to the questions of whether limiting choice decreases costs or has any effect on the quality of care (limiting choice to occupational medicine specialists would presumably increase the quality of care and reduce costs through prevention programs and more aggressive rehabilitation). Standardization of treatment through data-driven guidelines is another cost-control technique with the potential to improve treatment as well, given that most patients who are free to choose a provider are likely to be treated by a primary care provider with little or no training in occupational health and safety principles (Hashimoto, 1996). Case management, that is, the systematic coordination of medical, rehabilitative, and social services, assumes added importance in workers' compensation cases, in which medically unnecessary delays in return to work mean costly disability payments for the insurer and decreased income for most patients (Christian, in press). Note that under these circumstances, a narrow focus on cutting medical costs by reducing services may in fact increase overall costs by delaying return to work. Eccleston (1995) reports that case management is required in about half the states that regulate managed care. Yet another experiment at cost control has been 24-hour coverage—integration of workers' compensation and group health coverage by using the same resources for both personal and work-related health care. In theory, this approach has a number of advantages, including cost savings from consolidation of administration, elimination of "double-dipping," reduction in cost shifting, and consistency and continuity of care. In practice, fundamental differences in the two systems (for example, the high cost of disability compensation that can outweigh savings from less than aggressive medical treatment) have limited adoption of this sort of consolidation (Leone and O'Hara, 1998).

Several studies made systematic attempts to assess cost savings subsequent to changes in the manner in which health care was provided. In all cases, however, numerous changes were introduced simultaneously, which makes it impossible to judge their relative contributions to subsequent cost savings and other effects. A well-documented pilot project in the state of Washington changed the method of payment from fee for service to capitation and required workers to use the occupational medicine clinics of one of two large managed care organizations (Sparks and Feldstein, 1997). A similar study in Maryland (Green-McKenzie et al., 1998) looked at two cohorts of injured workers before and after the introduction of a managed care initiative involving an on-site case management team, a preferred provider network, and safety engineering and ergonomics programs. Both of these experiments showed large decreases in employer or insurer expenditures, both medical payments and disability payments. Quality of care was not directly assessed in either study, but the Washington study included telephone interviews with injured workers at 6 weeks and 6 months after injury. No significant differences between the intervention and control groups were seen in reports of pain, mental health, or physical functioning at either time, but at 6 weeks the workers receiving treatment through the managed care organizations were somewhat less satisfied with their treatment overall and their access to care. The Maryland study reports only that despite freedom to opt out at any time, 99 percent of the injured workers in the managed care group chose to stay within the system. Also of note is the fact that the number of claims was actually higher in the managed care cohort, so it is unlikely that the large cost savings were simply the result of a denial of claims.

Twenty-nine states currently have some type of managed care program for workers' compensation, costs to insurers and employers are down (Mont et al., 1999), and workers' compensation is once more a profitable sector for insurers (National Council on Compensation Insurance, as cited by *Consumer Reports*, 2000). Managed care is thus very likely to be a major part of workers' compensation for the foreseeable future.

IMPLICATIONS FOR EDUCATION AND TRAINING OF OCCUPATIONAL SAFETY AND HEALTH PROFESSIONALS

It is difficult to say with confidence that the market-driven changes in health care delivery will continue to evolve with cost reduction as its major theme. Signs of competition on the basis of quality, as well as increasing pressure on governments to intervene with "patient bills of rights" and more specific mandates like those that prohibit "drive-through" births, suggest that the pendulum may have started to swing away from cost cutting as a prime mover. Nevertheless, there are a number of features of U.S. health care today that are likely to affect the occupational safety and health workforce, primarily that segment dealing with the clinical care of workers, for some time to come.

First, the promise of population-based medicine and a corresponding emphasis on prevention certainly imply a favorable climate for occupational safety and health, but in practice this promise has, by and large, not been fulfilled. High rates of turnover of health plan members have undermined the assumed long-term savings achieved from the use of preventive measures like vaccinations, since the recipient will likely belong to a rival plan when the benefits are realized.

Second, the need to generate revenue and save money has led to an emphasis on seeing more patients—a trend reflected in the data in Chapter 2, which indicate that occupational medicine physicians are increasingly employed by clinics and less often by industry. The Association of American Medical Schools has frequently expressed the view that this increased emphasis on seeing patients is interfering with the conduct of the research and education missions at academic health centers. It may compete with the preventive aspects of occupational health as well.

Third, emphasis on primary care physicians, nurse practitioners, physician's assistants, and other health care professionals instead of specialist physicians may undermine attempts to increase the small numbers of board-certified occupational medicine specialists. As an earlier Institute of Medicine report (Institute of Medicine, 1988) pointed out, increasing the ability of primary care physicians in occupational medicine is vital for the health and safety of workers, but that task is itself dependent on a larger supply of occupational medicine specialists. A similar argument applies to the undeniable need for more coverage of occupational safety and health in physician's assistant and nurse practitioner training.

Fourth, the use of teams to deliver health care means that occupational safety and health training programs should expose students to the delivery of health care by teams.

Fifth, the growth of managed care means that more occupational safety and health services (including those paid for by workers' compensation) will be delivered in a managed care setting, so occupational safety and health students should be exposed to managed care during their training programs. Rivo et al. (1995) and Meyer et al. (1997) propose new curricula that can be used to better prepare physicians for practice in the managed care setting.

Sixth, occupational safety and health students should understand health care financing and the pressure to reduce health care costs and the likely impact of these on the quality of occupational health services.

7

Education and Training Programs

ABSTRACT. The committee used a variety of sources to assemble estimates of the annual number of master's-level graduates in the four core occupational safety and health disciplines. Twenty-nine U.S. schools offer such degrees in the occupational safety field, and they graduate about 300 students annually. This number is extremely low, given the incidence of workplace injuries, but the apparent acceptability of baccalaureate degrees in safety (about 600 graduates annually) by employers limits the demand for master's-level safety professionals. Less than 10 students per year are awarded doctoral degrees, a level low enough to threaten the future of academic departments of occupational safety. The committee's best estimate of the annual production of master's-level industrial hygienists is approximately 400, a volume probably consonant with employer demand in the industrial sector that has most commonly used them. Forty institutions offer occupational medicine residencies, and they annually produce about 90 graduates, a number that is probably insufficient for simple replacement of existing occupational medicine specialists. Attracting applicants is a large part of the problem, since the field draws heavily from established physicians, for whom return to full-time student status is not feasible. A similar situation exists in nursing, and 21 schools of nursing award only about 50 master's-level degrees in occupational health nursing each year.

Curricula in all four OSH disciplines are predominantly technical and science based, with a physical sciences/engineering emphasis in safety and industrial hygiene, and a biological, health, and programmatic emphasis in nursing and medicine. National Institute for Occupational Safety and Health (NIOSH) training programs provide grants totaling approximately $10 million per year in support of OSH professional education, resulting in 300 to 400 master's degrees (or completed residencies) each year. Occupational medicine has been the recipient of the most funding, reflecting the high cost of postgraduate specialist training for licensed physicians. Industrial hygiene has followed closely, with occupational health nursing receiving about 55 percent of the funding received by occupational medicine, and occupational safety receiving about one-third of the funding received by occupational medicine. Because large numbers of small businesses do not employ OSH professionals, worker and manager training by the Occupational Safety and Health Administration (OSHA) and others is also reviewed. No degrees are associated with this training, which takes many forms, from simple handouts and videotape viewings to 1 to 2 weeks of classroom and hands-on instruction. An exhaustive survey was not attempted, but it is clear that tens of thousands of hours of worker training is done, largely in response to OSHA mandates.

The committee concludes that current problems in the education and training of OSH professionals include lack of sufficient emphasis on injury prevention, which is reflected most clearly in the very small number of doctoral-level graduates in occupational safety, limited support for students doing research in departments other than those that grant OSH degrees, and an inability to attract physicians and nurses to formal academic training in OSH. An existing problem likely to be exacerbated by the many changes under way in the work environment is the narrow focus on OSH personnel who primarily serve large, fixed-site manufacturing industries. A potential problem in responding to these changes in the future workplace is a lack of training in a number of areas of increasing importance. These areas include behavioral health, work organization, communication (especially risk communication), management, team learning, workforce diversity, information systems, prevention interventions, and evaluation methods. The committee also concludes that worker health and safety training, although abundant, is of unknown quality and efficacy and that manager training is rarely demanded, offered, or requested.

Any consideration of the future OSH workforce must include an analysis of the educational "pipeline" as it exists today. This chapter presents the best available estimates of both the number of OSH-related degrees being granted in the United States today and brief summaries of typical curricula. Because of the committee's concerns about the many small businesses, now and in the future, that do not employ OSH professionals, the chapter also provides a brief review of some of the major sources of continuing education and training for workers and managers, with and without OSH-relevant degrees, who are responsible for worker health and safety.

RESEARCH TRAINING

The development of new knowledge and its timely application are as central to OSH as to all other fields of human endeavor. Because of the toll of illness and disease of occupational etiology, much of the funded research in the field has been of an applied nature, often associated with the toxicology, epidemiology, or control of exposures to particular chemical, physical, biological, and safety hazards. Historically, the budgets of NIOSH or the National Institute of Environmental Health Sciences (NIEHS) to fund occupationally oriented research have been very modest by National Institutes of Health (NIH) standards. In the last several years, NIOSH has been successful in increasing extramural research funding in OSH via collaboration with other federal agencies in its National Occupational Research Agenda activities. NIOSH and NIEHS, have also supported research students via the training grant mechanism with NIEHS focusing largely on the field of toxicology.

As noted in Chapter 2, research relevant to OSH is conducted in diverse settings in the academic, government, and private sectors. In some cases research is carried out from institutional bases identified with OSH and in many others, research is carried out from disciplinary units in, for example, biology, engineering, or psychology. In the academic setting, the numbers of students who carry out work relevant to OSH applications are sparse, as are opportunities for interdisciplinary cross-fertilization. Except for doctoral programs in the traditional OSH fields, this precludes eligibility for standard NIH-type categorical training grants. Hence, support for research students with OSH interests in other fields is largely through individual investigator-initiated research grants from NIOSH, NIEHS, or other NIH sources.

As has been documented elsewhere in this report, changes in the workforce (Chapter 3), workplace (Chapters 4 and 5), and the delivery of health care (Chapter 6) present new research challenges, many of which lie on the fringes of or are completely outside the traditional OSH disciplines. The intersection of the workers' compensation system with managed care, the ethical challenges of managing the increasing ability to determine genetic susceptibility to workplace chemicals (Frank, 1999), and the quantification of the risk of musculoskeletal injury from repetitive tasks are current research topics only dimly perceived a decade ago. All require the deployment of new competencies into OSH research, and all illustrate the need to recruit a broader array of students to study these issues. In common with most research activities, it is difficult to predict from which approach the practical benefits will arise.

OCCUPATIONAL SAFETY PROGRAMS

According to the 1998 to 1999 College and University Survey conducted by the American Society of Safety Engineers (ASSE), 32 U.S. schools offer programs leading to a bachelor of science degree in safety (American Society of Safety Engineers, 1999). Thirteen more programs at 12 additional schools are reported to offer bachelor of science degrees in more general fields (e.g., applied science, industrial systems) with a concentration or option in safety. Bachelor's degrees with a minor in safety, 2-year associate of arts and associate of science degrees in safety, and certificates in safety are offered by an additional 24 schools (universities, colleges, technical colleges, and community colleges). Graduate degrees in safety are offered by 31 U.S. universities. Twenty-nine of these schools offer master of arts, master of science, or master of public health degrees, and nine advertise programs that lead to a Ph.D. or doctorate in public health.

As noted in Chapter 2, only seven schools offer programs accredited by the Accreditation Board for Engineering and Technology (ABET). ABET offers accreditation of programs awarding master of science and bachelor of science degrees, but only six Bachelor of Science programs and four Master of Science programs have been accredited to date. ABET does not accredit doctoral programs in safety.

During the summer of 1999 ASSE attempted to collect further information on schools that offer safety-related degrees and solicited historical data on their graduates and faculty. Figure 7-1 shows the numbers and types of degrees awarded by the 54 responding schools (73 programs) since 1990.

The ASSE survey also asked the schools to provide some estimate of the proportion of graduates employed in safety positions. Over the last 5 years, these estimates were fairly steady at about 60 to 70 percent for graduates with associate degrees, 45 to 55 percent for graduates with baccalaureate degrees, and 100 percent for recipients of doctorates. Employment in safety positions for master's degree graduates has climbed steadily from 75 to 94 percent. These data, especially those for graduates with associates and bachelor's degrees, should not be taken as reliable estimates of demand, since it is not clear how many were actually seeking employment in the safety field upon graduation, but they do indicate that some caution is in order in estimating supply from the number of graduates.

Perhaps the most worrisome aspect of all of these data is the small number of Ph.D.s being awarded. In only one year (1991) did Ph.D. recipients number exceed eight. Furthermore, analysis of limited data provided to the committee by NIOSH grantees showed only one dissertation in the

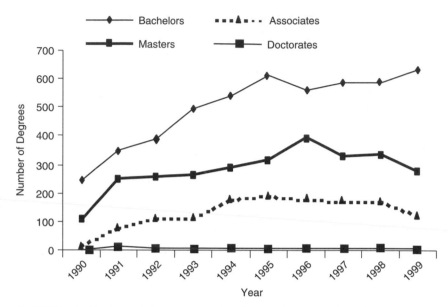

FIGURE 7-1 Number of degrees awarded from 1990 to 1999 by schools responding to ASSE survey on graduates. SOURCE: American Society of Safety Engineers (1999).

previous 5 years that focused on the traditional safety domain of acute traumatic injury prevention. Although few if any industries require safety professionals with doctorates, a critical mass of such individuals is necessary both for the conduct of critical research in injury prevention and for the continued viability of the academic programs that produce practicing safety professionals at the associate, bachelor's, and master's degree level.

Curricula

ASSE and the Board of Certified Safety Professionals (BCSP) have jointly published a series of curriculum standards that set forth the minimum academic requirements for both program accreditation and individual eligibility for attempting the Certified Safety Professional examination. The first of these (American Society of Safety Engineers and Board of Certified Safety Professionals, 1991) dealt with baccalaureate degrees, and subsequent efforts have addressed master's degrees in safety (American Society of Safety Engineers and Board of Certified Safety Professionals, 1994a), safety engineering master's degrees and safety engineering options in other engineering master's degrees (American Society of Safety Engineers and Board of Certified Safety Professionals, 1994b), and associate degrees in safety (American Society of Safety Engineers and Board of Certified Safety Professionals, 1995). No standard for doctoral degree programs has been published.

Baccalaureate Degrees

The standards for baccalaureate-level programs call for at least 120 semester hours of study, 60 of which must be upper-level courses. At least 54 hours of these upper-level courses must be in safety professional courses, and must include 34 hours in "professional core" courses (these are listed below).

Essential "university studies" courses for aspiring safety professionals include courses in six broad subject areas:

- mathematics and computer science (at least one semester each of calculus, statistics, and information processing);
- physical, chemical, and life sciences (two semesters each of physics and chemistry, one semester of human anatomy, physiology, or biology, and, if possible, a semester of organic chemistry);
- behavioral and social sciences and humanities (at least 15 hours, including a one-semester introduction to individual human behavior);
- management and organizational science (a one-semester introduction to business or management, and a course on business law or engineering law);
- communication and language arts (one-semester courses in each of rhetoric and composition, speech, and technical writing); and
- basic technology and industrial processes (applied mechanics, manufacturing processes).

"Professional core" courses are required to develop the basic knowledge and skills specific to the safety field. Subjects to be included are

- introduction to safety and health,
- safety and health program management,
- design of engineering hazard control,
- industrial hygiene and toxicology (often a series of courses),
- fire protection,
- ergonomics,
- environmental safety and health,
- system safety and other analytical methods, and
- an internship or cooperative course at an off-campus work site with a significant hazard control program.

"Required professional subjects" are important topics that may be covered by less than a full course:

- measurement of safety performance,
- accident and incident investigations,
- behavioral aspects of safety,
- product safety,
- construction safety, and
- education and training methods for safety.

The standards conclude with a list of safety elective courses in safety that range from those that are very industry specific (safety in the oil and gas industries) to those that are very general (public policy in safety and simulation and modeling in safety). The list is not definitive, however, and the only requirement is that the program include a number of electives, both general and safety related.

Master's Degrees in Safety

The authors of these standards recognize that there are two large groups of candidates for master's degrees: those who have undergraduate degrees in safety for whom the master's degree in safety is advanced study and those who have trained in a discipline other than safety for whom the master's degree in safety is initial preparation for a career in safety.

General criteria include a baccalaureate degree with a course structure very similar to that outlined above but allow courses in safety-related subjects such as risk management, industrial psychology, hazardous materials management, and quality control as substitutes for safety courses. Graduate study itself must include a minimum of 30 semester hours of study in a program significantly more specialized and advanced than baccalaureate programs. At least 20 hours must be devoted to safety science and safety professional practice, and the program should provide students with opportunities for participation in internships, field studies, research projects, and other interactive experience. A thesis is not strictly required, but programs without a thesis will be accredited only with a very convincing argument that it is impossible or impractical.

Specific course requirements are very similar to those for baccalaureate safety degrees described above for "professional core" courses.

Safety Engineering Master's Degrees or Engineering Master's with a Safety Option

Standards for safety engineering master's degrees are nearly identical to those for the master's degree in safety, except that the candidate must hold a baccalaureate degree from an ABET-accredited engineering program or must meet ABET equivalency requirements. To be recognized as a safety or safety engineering option (concentration or specialty are common synonyms) in another engineering master's degree program, a minimum of 9 semester hours in professional core courses must be included.

Associate Degrees in Safety

The standard for associates degrees in safety, which is primarily relevant to 2-year programs at community or technical colleges, is designed to help individual students prepare for either entry into the workforce as a safety technologist or technician or transfer to a 4-year accredited safety degree program. The standard thus has two tracks: one for *transfer* programs (i.e., for students who intend to go on to a 4-yr college for a baccalaureate degree in safety) and one for *terminal* programs (i.e., for students who intend on entering the safety workforce immediately).

Transfer programs give primary emphasis to building a solid foundation for upper level safety courses. The standard therefore calls for most of the university studies specified above for the baccalaureate degree standard and calls for only a limited array of safety courses (6 semester hours). The latter must include an introduction to safety and health, and the standard suggests fire protection, safety and health program management, and design of hazard controls as top choices for additional safety courses.

Terminal programs prepare students for entry-level positions as safety technicians or safety technologists. They emphasize applied knowledge of safety practices, safety laws and regulations, and accepted methods and procedures. The standard therefore calls for fewer hours of foundation subjects (math; physical, life, and behavioral sciences; business management; communication; and industrial processes) and more core and elective safety courses (27 semester hours). Included among the latter should be one industry-specific course (construction, transportation, forestry, chemistry, etc.).

Continuing Education

As noted in Chapter 2, certification is valid for only 5 years, at which point Certified Safety Professionals must provide BCSP with evidence of participation in professional development activities. Acceptable "continuation of certification" activities include membership in professional safety organizations, committee service for such organizations, and publications and papers but also include professional development, continuing education, and college courses. Although college courses must be taken at institutions accredited by the Council on Higher Education Accreditation, most safety-related professional development and continuing education courses sponsored by recognized organizations and many local, regional, and employer-sponsored conferences and courses are acceptable. Home study courses are acceptable if they award continuing education units or college credits or meet the standards of the Accrediting Commission of the Distance Education and Training Council. "Safety-related" is defined by BCSP as subjects that are included in the Certified Safety Professional examinations. No attempt was made to estimate the total number of such courses, or the number of students who take them each year, but some information of this sort is provided later in the chapter for several of the most prominent sources of such training, including NIOSH and other federal agencies. Additional information on distance education and training is presented in Chapter 8.

Future Needs

A major current need that is likely to become critical within the next decade if it is not remedied is the shortage of classically trained doctoral-level safety professionals. Of the 6.1 million nonfatal injuries and illnesses reported by employers in 1997, 5.7 million were injuries that resulted in lost work time, medical treatment beyond first aid, loss of consciousness, restriction of duties, or transfer to another job (Bureau of Labor Statistics, 1998a). However, as described in a subsequent section of this chapter, NIOSH funding for education of safety professionals is only about one-third of that for occupational medicine physicians and industrial hygienists and about two-thirds of that for occupational health nurses. Although the majority of safety professionals function quite well with a mixture of formal education at the baccalaureate or master's level, continuing education, and on-the-job experience, a continuing stream of such individuals depends upon a dwindling cadre of Ph.D. scientist-educators. Academic departments need a critical mass to attract both research funding and interested students, so special attention should be given to means of recruiting graduate-level faculty to teach and conduct research in this area. Possible approaches include grant support for regional occupational injury prevention centers similar to those supported by Centers for Disease Control and Prevention's National Center for [home and recreational] Injury Prevention and Control. The centers approach is especially attractive because of their inclusiveness: they can draw upon and support faculty and students from a wide variety of academic departments, an important attribute in a field as broad as safety.

INDUSTRIAL HYGIENE PROGRAMS

As indicated in Chapter 2, the most common academic degree for industrial hygienists is the master's degree, either a Master of Public Health degree with a specialization in industrial hygiene, or a Master of Science degree. The latter are often offered by departments of engineering, chemistry, environmental science, or other departments not readily identifiable as sources of industrial hygiene education. Perhaps for that reason estimates of the number of programs that offer degrees in industrial hygiene range from 40 to 50 (Peterson, 1999) to over 60 (Whitehead and West, 1997). The ASSE 1998 to 1999 College and University Survey identified 54 such institutions. What is known with certainty is that there are ABET-accredited baccalaureate programs at 5 U.S. institutions and ABET-accredited master's programs at 26 (Accreditation Board for Engineering and Technology, 1999). Of the latter, 21 programs receive NIOSH support. They have awarded an average of 210 master's degrees per year over the last 5 years. Including an estimate of graduates from nonaccredited programs yields an annual production of 400 to 600 master's-level industrial hygienists. A realistic estimate would be near the lower end of that range—400 per year. That would be consistent with the steady 300-per-year increase in Certified Industrial Hygienist (CIH) certificates awarded since 1980.

Curricula

The Related Accreditation Commission of ABET, with the assistance of the American Academy of Industrial Hygienists, provides a moderately detailed description of the required curricula for "industrial hygiene and similarly named engineering-related programs." Master's programs entail a minimum of 30 semester hours of interdisciplinary instruction and include special projects, research, and a thesis or internship. Research capability, management skills, and government relationships may be the subjects of special emphasis. A minimum of 18 semester hours must be devoted to industrial hygiene sciences and industrial hygiene practice. Epidemiology and biostatistics are examples of the former, which ABET defines as extensions of basic science or mathematics to industrial hygiene. Courses on industrial hygiene practice apply industrial hygiene sciences to specific societal needs and require open-form problem solving, cost and ethical considerations, and independent judgment in integrating specialty areas into professional service. Control of physical and chemical hazards, environmental health, and occupational safety are listed as typical topics. The remaining 12 "unspecified" hours allow further specialization. Common topics are public health, environmental law, and management techniques.

In practice, a typical master's program might involve the following courses:

- introduction to industrial hygiene,
- introduction to safety,
- introduction to occupational health,
- biostatistics and epidemiology,
- toxicology,
- chemical hazards,
- physical hazards (including radiation),
- exposure assessment (with an industrial hygiene laboratory),
- engineering controls,
- legal and regulatory issues in occupational health and safety,
- introduction to ergonomics,
- internship, and
- a project or thesis.

Continuing Education

As indicated in Chapter 2, continuing education plays a large role in the training of professional industrial hygienists. Not only is annual participation a requirement for maintaining certification but it has also been the primary means of entry into the field for those unable to devote several years to graduate education and the most common introduction to industrial hygiene for worker and management personnel with responsibility for health and safety in medium-sized to large firms. The American Industrial Hygiene Association, the American Conference of Government Industrial Hygienists, the NIOSH Education and Research Centers and other universities that offer degrees in industrial hygiene and a myriad of private firms offer American Board of Industrial Hygiene (ABIH)-approved courses. A full accounting is beyond the committee's resources, but in academic year 1996–1997, for example, programs supported by NIOSH alone offered a total of 234 courses to 5,621 people throughout the country.

Future Needs

Whitehead and West (1997) conducted the only quantitative study of future demand for ABIH-certified industrial hygienists and the ability of current training programs to meet that demand. Their predictions of demand involved application of the current industrial hygienist/worker ratios to the numbers of workers projected for each Standard Industrial Code in 2005 by the Bureau of Labor Statistics. A more elaborate model used a straight-line extrapolation of the upward trend in the industrial hygienist/worker ratio between 1989 and 1995 to the year 2005. Their conclusion was the same in both models: current training capacity is probably adequate to meet the requirements generated by the models. However, it is clear that the changes in work and the workforce documented earlier in this report (see Chapters 3 through 5) demand increased flexibility and innovation in the industrial hygiene curriculum, even for those destined for employment in traditional industrial hygiene positions with medium-sized to large industrial sector employers. Although the core competencies driving graduate training programs must continue to be based in the natural sciences, there are clear needs for these professionals to gain an understanding of:

* organization, operation, and management in the economic sectors in which they operate;
* the behavioral and psychosocial factors that affect worker health and safety; and
* methods for and approaches to the education and training of both workers and managers.

It is not clear that this material can simply be added to the existing curricula, given the constraints of time and accreditation standards, although there is a continuing need to reevaluate both curriculum content and accreditation requirements. An alternative means of encouraging a broader skill set is through modifications in the CIH certification examination or more structured continuing education requirements for certification maintenance.

The most important issue, however, was raised in Chapter 2, and that is that the need for industrial hygienists is much greater than the demand. This assertion is based, first, on the argument that because industrial hygienists and other OSH professionals are hired almost exclusively by medium-sized to large industries, much less than half of the current workforce ever interacts with an OSH professional of any sort, and probably less of the future workforce ever will do so. Second, even within the traditional areas of OSH practice, there has always been a tendency to integrate and merge technical and managerial functions, which can result in assigning industrial hygiene responsibilities to people with administrative and managerial skills instead of scientific and technical skills. The current trend toward outsourcing industrial hygiene services is another trend that separates the preventive aspects of the hygienist's role from both workers and workplace decision makers.

The issue facing educators over the next decade is therefore not how to produce more of the same or even similar numbers of slightly different industrial hygienists but how to continue providing the current level of supply while developing a new model of industrial hygiene practice appropriate for small, service-oriented, multisite businesses and their diverse and transient employees.

OCCUPATIONAL MEDICINE RESIDENCIES

As of January 1999, 40 institutions offered occupational medicine residency programs (American College of Preventive Medicine, 1999; Carol O'Neill, American College of Preventive Medicine, personal communication, June 4, 1999). These programs, which are distributed throughout the United States, provide 86 positions for the academic year and 95 positions for the practicum year. Some programs combine the academic and practicum years, but the net result is approximately 90 graduates per year. Data presented below in the section on NIOSH-supported programs indicate that NIOSH provides support to approximately 50 residents annually, so it is clear that such support plays a large role in ensuring the supply of occupational medicine physicians.

Some residencies are in schools of public health and others are in hospitals. The Institute of Medicine (IOM) has issued a report about incorporating the environmental and occupational medicine theme into the medical school curriculum (Pope and Rall, 1995; Goldman et al., 1999), but with less than 25 percent of all medical schools offering an occupational medicine residency program, such incorporation would appear to be difficult for the majority of medical schools in the United States. Medical schools without training programs are less likely to have faculty members knowledgeable in an area than are medical schools with such training programs. Further, not all of these 40 institutions offering occupational medicine residencies are medical schools.

A possible alternative to traditional residencies is suggested by an innovative distance learning program offered by the Medical College of Wisconsin, which provides the required academic program and a Master of Public Health (MPH) degree to physicians by means of a self-paced, computer-based curriculum that requires students on campus only for orientation and graduation. The program, which graduated its first students with an MPH degree in 1988, is almost entirely on-line. The mechanics of the program are discussed in the next chapter, but as of June 1999 the program had graduated 347 students (48 in June 1999) and had

250 students currently taking one of the ten 4-month courses required (Greaves, 1999).

Students from this program have a very favorable rate of passing the American Board of Preventive Medicine board examination. In 1997, for example, 92.5 percent of the graduates of the Wisconsin program passed the examination whereas 80 percent of conventional residency graduates and 61 percent for physicians without residency training who were eligible for the examination via the equivalency route passed the examination. In each of the past 3 years, approximately one-quarter of all physicians who have become board certified in occupational medicine had taken at least one course from the Medical College of Wisconsin Master of Public Health program.

Curricula

As noted in the Chapter 2, a residency entails successful completion of three separate phases of training: a clinical year, an academic year, and a practicum year. The following description of the content of those phases follows very closely that of the Residency Review Committee (RRC) of the Accreditation Council for Graduate Medical Education (ACGME).

The clinical phase, formerly called an "internship," constitutes a graduate year of clinical education and experience in direct patient care. It must provide broad experience in direct patient care, including ambulatory and inpatient hospital experience, for example, family practice, pediatrics, internal medicine, and obstetrics and gynecology, or in a transitional year; and it must be completed before the start of the practicum phase. The curriculum generally includes at least 3 months of hospital experience in internal medicine. A balanced program of outpatient and inpatient care, including pediatrics, obstetrics and gynecology, and emergency medicine, provides a broad range of exposure to other types of patients.

The academic phase consists of a course of study that leads to the Master of Public Health or equivalent degree in an institution accredited by a nationally recognized accreditation agency. The course of study requires a minimum of 1 full academic year or its equivalent, as determined by RRC. The resident who is in a degree-granting institution must meet all requirements for the degree. When no degree is granted, a postbaccalaureate curriculum of content, depth, and rigor equivalent to a degree program may be acceptable, provided that the curriculum is approved by an appropriate body, such as the graduate school or academic senate of an institution accredited by a nationally recognized accreditation agency. The required course work must include biostatistics, epidemiology, health services organization and administration, environmental and occupational health, and social and behavioral influences on health. In addition, adequate opportunities should be available for the resident to participate in research activities appropriate to the chosen field of education.

The practicum phase is a year of continued learning and supervised application of the knowledge, skills, and attitudes of preventive medicine in the field. The purpose of this phase is to prepare the resident for the comprehensive practice of occupational medicine and therefore must provide opportunities for the resident to deal with clinical, scientific, social, legal, and administrative issues from the perspectives of workers and their representatives, employers, and regulatory or legal authorities. For at least 4 months the resident engages in supervised practice within the real world of work. Through interaction with occupational health personnel, workers, human resources and industrial relations personnel, line supervisors, worker representatives, and the medical community, the resident must gain experience in the clinical and administrative aspects of direct worker care and job assignment, medical screening and surveillance, health conservation and promotion, environmental assessment, employee assistance, and relevant regulatory compliance.

Appropriate practicum opportunities may be found in a variety of settings. Suitable sites may include firms involved with heavy and light manufacturing; the utility, service, and transportation sectors; and clinics that provide comprehensive services to workers and employers. Although diagnostic and referral clinics that specialize in occupational disease can have a vital role in the education of residents, they do not afford the broad practicum opportunities specified above. This experience need not be obtained at a single site, nor must the experience be received in a single 4-month block. However, sufficient sustained attendance at each facility must occur to permit the assumption of significant clinical and administrative responsibility.

The content of the practicum phase is supposed to offer clinical, administrative, and didactic components in settings in which there are well-organized programs appropriate to the particular field of preventive medicine and supervised application of the knowledge, skills, and attitudes of preventive medicine gained in the academic phase and in the didactic component of the practicum phase.

A didactic component can include structured lectures, journal clubs, symposia, and other activities that focus on:

1. physiological responses to heat, pressure, noise, and other physical stresses;
2. the diagnosis, prevention, and treatment of occupational disease;
3. toxicology and risk assessment;

4. industrial hygiene instrumentation and basic environmental control measures;

5. behavioral factors in accident causation and control and medical support of accident investigations;

6. ergonomics;

7. determining fitness to work, placement of workers, and adaptations of work to accommodate handicaps;

8. employee assistance programs;

9. health education and health promotion; and

10. occupational health data management and analysis.

The clinical component shall include but not be limited to:

1. clinical care of workers in the prevention, diagnosis, treatment, and rehabilitation of work-related disorders;

2. evaluation of the fitness of workers for normal or modified job assignments in a wide variety of work environments and the assessment of impairment and disability; and

3. counseling and education of workers and supervisors with respect to work or environmental hazards, health-related habits of living, and the arrangements of work.

Finally, an administrative component shall provide the resident with opportunities for management responsibilities and shall include but not be limited to each of the following topics:

1. the planning, administration, supervision, and evaluation of a broad program for the protection and promotion of the health and safety of workers in the work setting, including health risk assessment, accident evaluation, and risk reduction;

2. application of administrative and scientific principles to achieve compliance with regulatory requirements and the requirements of workers' compensation plans; and

3. acquisition, collation, storage, and analysis of health and environmental data.

Continuing Education

Occupational medicine physicians, like all physicians, are required to participate in a certain amount of continuing medical education to keep their medical licenses. In addition, as described earlier in Chapter 2, physicians who become board certified in occupational medicine in or after

1998 must be recertified every 10 years, in part on the basis of continuing education credits.

During 1998 the American College of Occupational and Environmental Medicine (ACOEM) sponsored 26 educational opportunities that reached more than 4,200 participants, and jointly sponsored more than 100 other educational activities, live and via distance leaning (American College of Occupational and Environmental Medicine, 1999b). ACOEM's two conferences each year—in the spring and fall—provide courses, classes, and presentations at which physicians can obtain continuing education credits. Typical offerings are full-day postgraduate seminars on such topics as myofascial pain syndrome, silica toxicology, occupational skin diseases, and travel medicine, along with medicolegal testimony and occupational medicine self-assessment review. ACOEM also offers professional development courses at other locations and dates. The 1999 courses were Impairment and Disability Evaluation, Medical Review Officer Training, and Occupational Medicine Board Review. The component organizations (regional occupational medicine organizations) also provide continuing education activities with annual or periodic conferences. A survey of ACOEM members (The Gary Siegal Organization, Inc., 1996) found that 75 percent of respondents indicated that they had received some funding for continuing education in 1996. Of those who did, 32 percent received $1,500 or less, 44 percent received between $1,500 and $3,000, and 24 percent received more than $3,000.

Many of the university programs provide 1- to 3-hour long continuing education activities at weekly, fortnightly, or monthly conferences. Universities also offer a regular catalog of 1–3 day conferences or courses on specific areas. Among these varied venues a sufficient number of continuing education activities appear to be available for all physicians both to keep them current and to fulfill their requirements for licensure and recertification.

Future Needs

According to a former president of ACOEM (Anstadt, 1999), the number of graduates from residency programs is insufficient to meet the current and future demands for occupational medicine physicians. The supply is below the numbers needed for replacement of existing occupational medicine physicians. The occupational medicine residencies report difficulties in the following areas:

- attracting sufficient number of applicants,
- funding the residents, especially for the academic year, and
- having sufficient numbers in some programs to constitute a critical mass.

The barriers to attracting potential residents to residencies include the following:

• Residencies cost institutions $30,000 to $40,000 per year, and unlike more clinically oriented residency programs, occupational medicine residents have relatively little direct patient contact to generate funds.

• Physicians who anticipate doing nearly 100 percent outpatient clinical activities (the recent trend in occupational medicine) see less value in occupational medicine residency; thus, the supply of future residents may actually decrease.

• Most importantly, many physicians become interested in occupational medicine some years after graduation from medical school. Pearson and colleagues (1988) surveyed a random sample of preventive medicine specialists and found that of 166 self-identified occupational medicine specialists, 86 percent had entered the field after one or two career changes and 51 percent were 44 years of age or older. Return to full-time student status in an occupational medicine residency is not an attractive option for this population.

The Occupational Physicians Scholarship Fund, an independent organization founded and directed by occupational medicine physicians from both academia and a wide variety of industries, has been successful in providing funding for some residents. However, because of financial limitations, the number of residents that can be supported is limited. To date the fund has provided 122 scholarships to 84 individuals. By the year 2011 the total number of scholarships is expected to be 250. The fund averages 45 applicants per year. Clearly, not all applicants are able to receive funding (Bronstein, 1998). An attractive option is the more widespread use of distance education like that conducted by the Medical College of Wisconsin. This is explored further in Chapter 8, but it addresses only the academic year of residency. The practicum year remains a serious obstacle for most established physicians. A more comprehensive solution requires reexamination of the current pathways to certification in occupational medicine, a major recommendation of the 1991 IOM Committee Addressing the Physician Shortage in Occupational and Environmental Medicine (Institute of Medicine, 1991). That study suggested exploring the possibility of offering certificates of added qualification to physicians who are board certified in internal medicine or family practice and who also have advanced training or experience in occupational medicine (a certificate for added qualification in geriatric medicine is already available). The IOM study also suggested modifying the current pathway to dual certification (internal medicine and occupational medicine) by allowing a year of occupational medicine practice in place of a practicum year. The current IOM committee believes that these ideas should be explored again. An even simpler alternative would be for the American Board of Preventive Medicine to eliminate the restriction of its existing equivalency (experience) pathway to physicians who graduated before 1985.

OCCUPATIONAL HEALTH NURSING PROGRAMS

As noted in Chapter 2, occupational health nursing programs are a specialty focus generally offered at the graduate level. The American Association of Occupational Health Nurses has identified a total of 21 programs that offer graduate degrees in occupational health nursing. Together they produce about 50 graduates annually. This is accomplished primarily through funding from NIOSH Education and Research Centers and Training Program Grants. These programs may be in schools of public health or schools of nursing offering both master's and doctoral degree programs. In addition, a few non-NIOSH-funded schools of nursing across the country offer occupational health nursing graduate programs.

Accreditation for all of these programs is provided by the National League for Nursing (NLN) or the Council on Education in Public Health (CEPH). Occupational health nurses work in industry, manage clinics, develop corporate occupational policies, and design and implement health promotion, disease prevention, and health surveillance programs for employees. Although there is no set standard for a program of study for occupational health nursing, the programs offered follow guides set forth by NIOSH and the school discipline accreditation bodies (i.e., NLN and CEPH), with an emphasis on interdisciplinary learning.

Curricula

In general nurses receiving a master's degree in public health with an occupational health nursing focus take courses with content in public health sciences, occupational health sciences, and occupational health nursing and often have a practicum experience. The public health science course work includes epidemiology, biostatistics, environmental sciences, health administration, and behavioral sciences. Epidemiology provides a foundation for recognizing trends in occupational illness and injury; biostatistics is important in the planning, coordination, and analysis of research; course work in environmental sciences introduces basic concepts of environmental health and recognition of environmental hazards (e.g., air and water pollution, food safety, hazardous substance exposure, and environmental policy and management); course work in health policy and administration focuses on organizational and human resources management; and course work in behavioral science provides an understanding of human behavior and motivational theory and how they relate to health choices. Course work in the occupational health sciences typically covers safety and ergonomics, occupational health, toxicology, and industrial hygiene. Each of these courses provides an introduction to the basic concepts and principles of each discipline. In addition, a framework for interdisciplinary collaboration is enhanced through joint field experiences, workplace walkthroughs, and occupational health problem-solving assignments and seminars. The goal of occupational health nursing content is to have a better understanding of means of assessing of worker and workforce illnesses and injuries as well as a better understanding of health promotion and protection concepts and principles so that effective intervention and prevention strategies can be designed and evaluated to improve worker health and safety and working conditions. Role function and leadership concepts and principles are emphasized.

Field praticum experiences that offer learning opportunities not available in the classroom may be provided. The purpose and potential benefits are to relate theoretical classroom learning to practice situations; gain experience, skills, and confidence in dealing with administrative, educational, and service problems; explore and have an increased understanding of the structure and dynamics (e.g., agency objectives, goals, values, resources, and constraints) of a specific work setting; and identify problems for intervention, prevention, and evaluation. Completion of a master's paper specific to occupational health and safety and, often, a comprehensive examination of the subject matter conclude the program of study.

A master's degree in nursing with an occupational health focus is another degree option. It can often have either a management and administration focus or a nurse practitioner focus. The management and administration focus offers course work in nursing administration, financial management, informatics, and management principles applied to the occupational health setting. The nurse practitioner focus offers course work on the theory and practice of adult health maintenance and the assessment and management of common ailments facing working adults. Clinical residencies in specialty clinics and occupational health settings with experts in the field are provided.

In addition to course work in the occupational health sciences, course work may include such topics as advanced physiology and pathology, pharmacology and therapeutics, diagnostic processes, occupational health management principles, and primary care. Practicum experiences relevant to the specialty and nursing role function, research, and a thesis may also be included.

Several programs are offered in occupational health nursing at the doctoral level. Preparation is designed to prepare nurse researchers who can design studies to address researchable problems in occupational health and safety.

More recently, there has been increasing emphasis in offering graduate occupational health nursing programs in nontraditional formats, such as through distance learning or weekend education. With technological advancements occurring on a daily basis, this type of educational format will grow. These options allow students to take course work through the Internet, or independent study with limited on-campus time or through programs of study offered on campus but in a periodic weekend format.

Continuing Education

Mandatory nursing continuing education varies with state. However to be eligible for optional certification in occupational health nursing and renewal of that certification in 5-year increments, a minimum of 75 continuing education contact hours must be posted within the 5-year period prior to certification and recertification.

Occupational health nurses can obtain continuing education from numerous sources. The professional association for occupational health nurses, AAOHN and its 180 chapters across the nation, provide ample avenues for attainment of continuing education. For example, at the 1999 American Occupational Health Conference, AAOHN offered more than 130 continuing education courses for participants. In addition, universities, particularly those with occupational health and safety programs such as the NIOSH Education and Research Centers, provide seminars, workshops, conferences, and courses to meet the learning needs of the professional occupational health nurse.

Future Needs

The major shortfall in the occupational health nursing field is the same as that of the occupational medicine field; that is, there is not so much a shortage of practitioners as there is a shortage of practitioners with formal training in the field. The barriers to the production of more master's-level occupational health nurses are also similar to those that inhibit potential occupational medicine residents, especially the reluctance of mid-career professionals to return to full-time student status for 2 years. As with occupational medicine residents, the key to increasing the supply of master's-level occupational health nurses lies in more accessible educational programs and financial assistance to achieve degree completion.

21st CENTURY KNOWLEDGE AND EXPERTISE

As is evident from the preceding review of typical curricula in each of the traditional OSH professions, the curriculum for each profession is firmly anchored in the natural sciences, provides a challenging educational experience, and for those who complete it, provides a solid technical preparation for entry-level OSH professionals. Nevertheless, as noted in the section on industrial hygienists, the many changes in the world of the U.S. worker documented in earlier chapters of this report appear to demand knowledge and expertise from all OSH professionals in still more areas. Behavioral health, work organization, communication (especially risk communication), management, team learning, workforce diversity, information systems, prevention interventions, and evaluation methods are some of the most important. To this list one might add methods for effective training of adult workers, the physical and psychological vulnerabilities of members of the workforce by age, gender, and socioeconomic and cultural background; the resources available for help with accident prevention and analysis; business economics and values; health promotion and disease prevention; community and environmental concerns, and the ethical implications of technological advances such as the mapping of the human genome (as well as the dual responsibilities of OSH professionals to workers and employers generally). Physicians, nurses, and other OSH healthcare professionals must be exposed to managed care and team health care delivery during their training, and need a solid grasp of health care financing and its effects on workers' compensation if they are to maximize their benefit to workers and employers. The use of multidisciplinary teams is one way of providing for all the requisite knowledge, skills, and abilities, but individual professionals must be familiar enough with these areas to know when and where to turn for help.

Simply adding courses in these areas to the curriculum is not feasible because adding months or years to any of these programs is likely to have a strong negative effect on the attractiveness of the programs to prospective students. Substitutions for areas currently in the curriculum will also be difficult because accreditation requirements dictate the current curricula and credentialing examinations reflect much of the current curricula. Academic accreditation bodies should periodically conduct comprehensive evaluations of the training programs that they judge, including empirical measures of graduates' satisfaction and success, paying special attention to the emerging needs listed here, and altering requirements (or standards) accordingly. Credentialing bodies, which generally dictate continuing education requirements, could use that mechanism to ensure that their technically trained members stay truly current as well.

FUNDING SOURCES FOR OCCUPATIONAL SAFETY AND HEALTH EDUCATION AND TRAINING

The remainder of this chapter describes the current sources, nature, and extent of support for OSH education and training in both the public sector and, less comprehensively, the private sector, beginning with NIOSH, the major source of funding for graduate education in the traditional OSH professions. Because of the committee's previously noted concern for the large segment of the American workforce that does not typically come in contact with these OSH professionals, subsequent sections describe current sources of funding for continuing education and worker and employer training.

NIOSH Education and Training Programs

NIOSH awards 1- to 5-year training program grants (TPG) and Education and Research Center (ERC) training grants to support educational programs in the fields of industrial hygiene, occupational health nursing, occupational medicine, occupational safety, and other specialized OSH training areas. The objective of this grant program is to award funds to eligible institutions or agencies to assist in providing an adequate supply of qualified professional and paraprofessional OSH personnel to carry out the purposes of the Occupational Safety and Health Act. Any public or private educational or training agency or institution that has demonstrated competency in the occupational safety and health field and that is located in a state, the District of Columbia, or U.S. Territory is eligible to apply for a training grant. NIOSH supports both short-term continuing education courses for OSH professionals and others with worker safety and health responsibilities and also academic degree programs and postgraduate research opportunities in the areas of occupational medicine, occupational health nursing, industrial hygiene, and occupational safety. Appendix D contains a list of the 1999 grant holders, and Figure 7-2 displays the funding for each of these four disciplines between 1995 and 1999. The "other OSH" category includes graduates from academic departments with dual names, e.g., Industrial Hygiene and Safety, Occupational Health and Safety, and Environmental and Occupational Health, and occasional graduates in related fields such as epidemiology and toxicology. (See Chapter 2 for a brief description of a new program to stimulate curriculum development on occupational health psychology.) Occupational medicine has been the recipient of the most funding, reflecting the high cost of postgraduate specialist training for licensed physicians. Industrial hygiene has followed closely, with occupational health nursing, at about 55 percent of occupational medicine, and occupational safety,

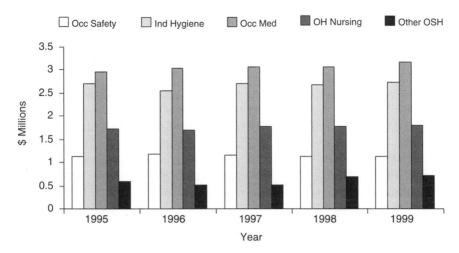

FIGURE 7-2 Funding, by discipline, by NIOSH of ERCs and TPGs, 1995 to 1999. Abbreviations: Occ, occupational; Ind, industrial; OH, occupational health; OSH, occupational safety and health. SOURCE: Susan Board, National Institute for Occupational Safety and Health, personal communication, November 29, 1999.

receiving about one-third of the amount received by occupational medicine. This pattern has been very steady over the 5 years for which the committee has data. The following sections provide information on the products of this spending.

Degree Programs

OSH programs take a variety of forms across the myriad of industries and businesses of the United States, and the educational backgrounds of those who supervise, conduct, and participate in them vary widely as well. This chapter, like Chapter 2, focuses on the four traditional OSH disciplines—occupational medicine, occupational health nursing, industrial hygiene, and occupational safety. As noted in Chapter 2, a large proportion of professionals in these fields have formal education beyond the baccalaureate. A master's degree is common for practitioners, and a doctorate is almost a requirement for teaching and research (in the case of occupational medicine, these degrees follow receipt of a medical degree and completion of a clinical residency). Figure 7-3 shows the number of master's degrees awarded in each of the traditional OSH fields by NIOSH-supported programs for each year between 1987 and 1997. The totals include graduates supported by either TPGs or ERC grants.

Industrial hygienists have clearly been the leading recipients of

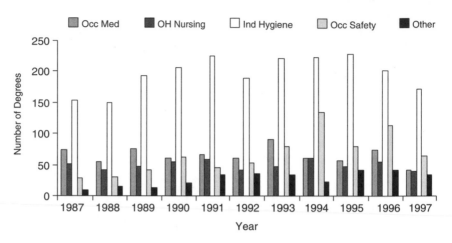

FIGURE 7-3 Master's degrees awarded with support of NIOSH training programs, 1987 to 1997. Occ, occupational. SOURCES: National Institute for Occupational Safety and Health (1988, 1989, 1990, 1991, 1992, 1993, 1994, 1995, 1996, 1997a, 1998b).

NIOSH support for master's-level education during the past decade, but Figure 7-3 does not fully represent the extent of NIOSH support for physicians. Most, if not all, of the master's degrees in occupational medicine represented in Figure 7-3 were awarded to physicians who were completing a residency, but not all institutions offering residencies in occupational medicine provide a degree as part of the process. The height of the occupational medicine bars in Figure 7-3 would be roughly doubled if physicians who successfully completed an occupational medicine residency that did not provide a master's degree were included. Significantly fewer graduates come out of NIOSH-supported programs in Occupational Safety (about 70 annually) and Occupational Health Nursing (a little less than 50 per year).

As noted in the discussion of Figure 7-2, the "other OSH" category includes academic departments with dual names, for example, industrial hygiene and safety, occupational health and safety, and environmental and occupational health, as well as related fields such as epidemiology and toxicology. NIOSH support for the "other" category is also underrepresented in Figure 7-3, in that this is the only one of the five categories for which NIOSH provides support for undergraduate education. In fact, the vast majority of NIOSH-supported graduates in the "other" category are undergraduates receiving 2-year associate degrees or 4-year bachelor's degrees (see Figure 7-4). A small number of additional students receive certificates, documenting the completion of approximately 30 credit hours in occupational safety and health.

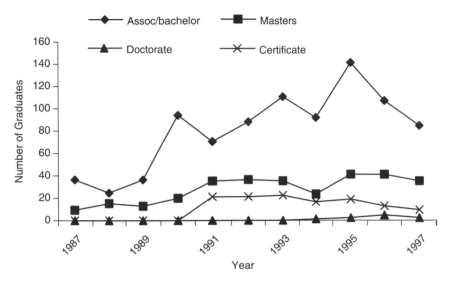

FIGURE 7-4 Graduates of "other" NIOSH-supported educational programs, 1987 to 1997. Assoc, associate degree. SOURCES: National Institute for Occupational Safety and Health (1988, 1989, 1990, 1991, 1992, 1993, 1994, 1995, 1996, 1997a, 1998b).

Continuing Education

In addition to formal degree programs, a wide variety of short courses on specific OSH topics is provided by colleges and universities affiliated with the NIOSH ERCs. Topics may be very general and widely applicable or may be specific to particular industries or businesses, and courses may be as short as a few hours to as long as a week or more. Many institutions provide instruction on the Internet or use other computer applications and off-campus sites to make the courses attractive and accessible to full-time workers. These courses are not associated with a degree program and do not generally demand prerequisites. As is the case with short-term courses offered by OSHA and the Mine Safety and Health Administration (MSHA), trade unions, industry, and professional organizations, there is no universally recognized accreditation body and, hence, no guarantee of quality, nor is there generally any attempt to assess the efficacies of these courses.

Figure 7-5 shows the number and type of continuing education courses and student-days of continuing education instruction provided by NIOSH-supported ERCs and TPGs. Number of student-days is simply the sum of the product of the number of students and the duration of their course.

Students who attend these courses are most often employed by pri-

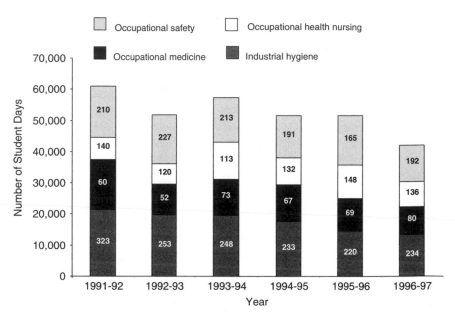

FIGURE 7-5 Student-days of continuing education and number of courses (inside the columns of the histogram) supported by NIOSH, 1991 to 1997. SOURCES: National Institute for Occupational Safety and Health (1988, 1989, 1990, 1991, 1992, 1993, 1994, 1995, 1996, 1997a, 1998b).

vate industry, but government at all levels and academic institutions send substantial numbers of employees as well. Figure 7-6 shows the categories of the employers of the 33,884 students who took NIOSH-supported continuing education courses in 1996-1997. The distributions in the previous years were very similar to those in 1996–1997.

Not all of the students who took these continuing education courses were traditional OSH professionals. In fact, as Figure 7-7 shows, only about half the students who took NIOSH-supported continuing education courses in academic year 1996–1997 were safety professionals, industrial hygienists, occupational physicians, or occupational health nurses. A small number were clearly identified as paraprofessionals, that is, trained aides who assist a professional person, but about half were simply classified as "other." Some of these students were certainly professionals from other OSH-related disciplines, but a large proportion was no doubt composed of managers, supervisors, and workers with some responsibility for providing or carrying out OSH programs and worker training at their workplaces.

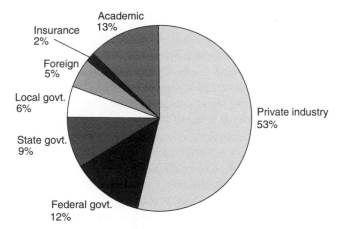

FIGURE 7-6 Employers of students attending NIOSH-supported continuing education courses in 1996–1997. Govt., government. SOURCE: NIOSH (1998b).

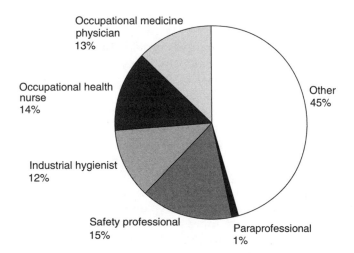

FIGURE 7-7 Backgrounds of students attending NIOSH-supported continuing education courses in 1996–1997. SOURCE: NIOSH (1998b).

OSHA-Supported Education and Training

Since the passage of the Occupational Safety and Health Act in 1970, NIOSH has been the primary federal agency supporting education and training of OSH *professionals*, but OSHA assumed responsibility for *worker* training and more than 100 OSHA regulations (standards) require worker training. Other OSHA standards make it the employer's responsibility to limit certain job assignments to individuals who are "certified," "competent," or "qualified," presumably by virtue of special previous training in or out of the workplace. A number of standards are quite explicit about what safe practices should be taught, but most are very general ("Methods shall be devised to train operators of powered industrial trucks"). OSHA has provided some voluntary guidelines to assist employers, but it has generally left the content and methods to employers.

A large number of private training entities offer training packages and videotapes to help employers meet these requirements, and OSHA itself has developed more than 80 short courses that it offers through its own facilities and that it makes available for use by others. The OSHA Office of Training and Education offers training and training programs to federal and state OSHA personnel, state consultants, other federal agency personnel, and private-sector employers and employees. This is accomplished through short-term training courses at the OSHA Training Institute in Illinois and 12 regional OSHA Training Institute Education Centers, the Susan Harwood Training Grants Program, and the train-the-trainer Outreach Grants Program.

OSHA Training Institute

OSHA Training Institute courses provide basic and advanced training and education in safety and health. Course subject matter emphasizes OSHA policies and standards as well as hazard recognition and hazard abatement techniques. Courses are designed to build a more effective workforce and to aid in professional development. The schedule contains courses for federal and state compliance officers, state consultants, other federal agency personnel, and private-sector employers, employees, and their representatives.

Approximately 80 different courses are offered, including 7-day overviews of OSHA standards for physicians and nurses and a 14-day course on criminal investigations for OSHA compliance officers. The modal course length is 4 to 5 days. More than 700 courses were taught in fiscal year 1998, and Figure 7-8 shows the extent of this instruction for the last 4 years for which data were available.

Continuing education units (CEUs) are available to participants in all

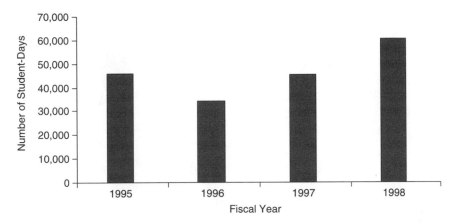

FIGURE 7-8 Student-days of instruction by OSHA Training Institute, all sites, in fiscal years 1995 to 1998. SOURCES: Office of Training and Education (1996, 1997, 1998).

courses conducted by OSHA Training Institute staff in accordance with the administrative and program criteria guidelines that have been established by the International Association for Continuing Education and Training. These CEUs also meet the criteria of the American Board for Occupational Health Nurses (ABOHN) for certification or recertification. Certification maintenance points are available to certified industrial hygienists who complete courses awarded points under the ABIH maintenance of certification program. All technical safety and health courses except the Collateral Duty Course for Other Federal Agencies meet Board of Certified Safety Professionals criteria for continuance of certification credit.

OSHA Training Institute Education Centers

In recent years the demand for OSHA Training Institute courses from the private sector and from other federal agencies has increased beyond the capabilities of the OSHA Training Institute. To address this increased demand for its courses, the OSHA Training Institute has established regional OSHA Training Institute Education Centers. These centers are nonprofit organizations that were selected after competitions for participation in the program. There are currently 12 centers nationwide, and each center offers 12 core OSHA Training Institute courses, generally on campus, but sometimes at other sites when demand is sufficient. Figure 7-9 shows that they clearly reach an audience different from that reached by the parent institute.

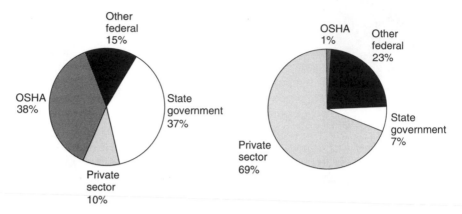

FIGURE 7-9 Source of fiscal year 1998 students for OSHA Training Institute (left) and OSHA Training Institute Education Centers (right). SOURCE: Office of Training and Education (1998).

Susan Harwood Training Grants Program

OSHA began a serious effort to increase and improve worker training in the late 1970s with a program called New Directions in which several million dollars in grants were given to unions and trade associations to develop training materials for workers and to provide training. A very large volume of training materials and training were generated from those initial grants. Many safety and health programs were started at unions and trade associations as a consequence of this seed money. The New Directions program ended in the early 1980s and was subsequently replaced by more targeted grant programs (e.g., programs targeted to certain hazards or hazardous industries) such as the Susan Harwood Training Grants Program.

The Susan Harwood Training Grants Program provides funds to nonprofit organizations to train workers and their employers to recognize and prevent safety and health hazards in their workplaces. It emphasizes three general areas: businesses with less than 250 employees, new OSHA standards, and OSHA-defined high-risk activities. Organizations that receive grants are expected to develop training programs that address a topic named by OSHA, recruit workers and employers for the training, and conduct the training. Awards have been made to safety and health organizations, employer associations, labor groups, educational institutions, and other nonprofit organizations. The competitive grants are for 12 months, although they may be renewed, and average about $100,000 (the grantees must contribute a minimum of 20 percent of the total project

TABLE 7-1 Subject Area and Number of Susan Harwood Training Grants Program Grants Awarded in Fiscal Year 1999

Subject Area	No. of Grants
Construction	7
Ergonomics	8
Food processing	4
Logging	3
Scaffolding	4
Shipyard safety and health	5
Silica in general industry	2
Small businesses	9
Worker outreach	4
Workplace violence	2

SOURCE: Occupational Safety and Health Administration (1999a).

cost). The topics for grants can be broad ones that cut across many occupations or extremely narrow ones that focus on one kind of injury in one occupation, and they are specified each year in the application announcement. The subject areas for fiscal year 2000 awards are prevention of amputations in manufacturing, hazards in health services facilities, and prevention of falls by construction workers. Table 7-1 shows the number of awards made in each of the 10 fiscal year 1999 subject areas.

A grant to the Oil, Chemical and Atomic Workers International Union in the area of worker outreach is typical. The union received $151,000 to develop a curriculum covering worker rights under the OSHAct and whistleblower protections administered by OSHA. It will instruct trainers who will then train local union members.

Outreach Grants Program

The OSHA Outreach Grants Program is a train-the-trainer program that authorizes individuals who have completed a 1-week OSHA training course to teach 10- or 30-hour courses in general industry or construction safety and health standards. These authorized trainers are also provided with OSHA course completion cards to give to their students. These cards not only provide workers with portable proof of training but also provide OSHA a means of tracking the amount of training taking place.

There is one course for the construction industry, course 500, and one course for general industry, course 501. These 1-week courses cover material that would-be trainers will subsequently cover in 10- or 30-hr courses for workers and concentrate on OSHA standards. Prerequisites for enrolling in course 500 (construction industry) are 5 years of construction safety experience and completion of course 510, Occupational Safety and Health Standards for the Construction Industry. The courses are conducted by the OSHA Training Institute and by 12 OSHA Training Institute Education Centers located around the country. Completion of the course authorizes the student to train others for 4 years, after which an update course is required to renew the authorization to provide training for another 4 years.

Subsequent training conducted by graduates of course 500 or 501 is a mixture of mandatory topics and instructor-chosen materials specific to a particular industry or work site, and is up to 10 hours for entry-level participants or 30 hours for students seeking broader of more intensive coverage. In 1998, more than 8,700 such classes were held and more than 150,000 student cards were issued to workers as evidence of course completion (Occupational Safety and Health Administration, 1999b).

State Training and Education Initiatives

Section 18 of the Occupational Safety and Health Act of 1970 encourages states to develop and operate their own job safety and health programs. OSHA approves and monitors state plans and provides up to 50 percent of an approved plan's operating costs. States must set job safety and health standards that are "at least as effective as" comparable federal standards. (Most states adopt standards identical to federal ones.) States have the option to promulgate standards that cover hazards not addressed by federal standards.

A state must conduct inspections to enforce its standards, cover public (state and local government) employees, and operate OSH training and education programs (Occupational Safety and Health Administration, 1999c). In addition, most states provide free on-site consultation to help employers identify and correct workplace hazards. At present, 23 states and jurisdictions operate complete state plans (which cover both the private sector and state and local government employees), and two states, Connecticut and New York, cover public employees only.

According to the annual report of the Occupational Safety and Health State Plan Association (1999), in fiscal year 1998, states with state plans provided 12,344 training programs for more than 262,000 employers and employees. Topics included confined spaces, communication about hazards of workplace chemicals, excavation safety, bloodborne pathogens, ergonomic hazards, and violence in the workplace. Many states are following OSHA's lead in making use of the Internet to provide a wealth of OSH information to anyone with computer access. Oregon has gone one step further and now offers five core workshops and two advanced work-shops entirely via the Internet. Students take the classes electronically, respond to questions, receive personal attention from a trainer, and receive a certificate of completion at terminals in homes, workplaces, schools, and libraries anywhere in the state.

NIEHS-Supported Training and Education

NIEHS was given responsibility for initiating a training grants program under the Superfund Amendments and Reauthorization Act of 1986. Although this training is focused on only a few occupations, albeit very hazardous ones, many of the programs developed have been considered among the very best, not simply in content but also in their attention to delivery and posttraining evaluation.

Since initiation of its Superfund Worker Training Grants Program in 1987, NIEHS has been funding non-profit organizations with a demonstrated track record of providing OSH education to develop training for and deliver training to workers involved in handling hazardous waste or in responding to emergency releases of hazardous materials. The major objective of the NIEHS Worker Education and Training Program is to prevent work-related harm by assisting in the training of workers on how best to protect themselves and their communities from exposure to hazardous materials encountered during hazardous waste operations, hazardous materials transportation, environmental restoration of nuclear weapons facilities, or chemical emergency response. During the first 8 years of the Superfund Worker Training Program (fiscal years 1987 to 1995), NIEHS has successfully supported 20 primary awardees. These represent more than 90 different institutions that have trained more than 500,000 workers across the country and presented nearly 25,000 classroom and hands-on training courses, which have accounted for more than 8 million contact hours of actual training. In 1995, eighteen awardees, in conjunction with more than 70 collaborating institutions, delivered 5,348 courses, that reached 87,205 workers. The courses ranged from 4-hour refresher programs through more complex train-the-trainer courses that lasted up to 120 hours.

Funding for the development or expansion of worker training programs from NIEHS depends on reauthorization of Superfund legislation, but through interagency agreements with the Environmental Protection Agency (EPA) and the U.S. Department of Energy (DOE), NIEHS continues to support the development and delivery of model worker health and safety training in three areas: hazardous waste worker training (HWWT), DOE nuclear weapons cleanup training, and the minority worker training program (MWTP). The National Clearinghouse for Worker Safety and

Health Training for Hazardous Materials, Waste Operations, and Emergency Response provides information and communication services for the awardees in these areas (National Clearinghouse for Worker Safety and Health Training, 1999). A brief summary of their activities in the year ending August 31, 1998, is as follows:

• HWWT—The 17 primary worker training awardees, in conjunction with more than 80 collaborating institutions, delivered 4,820 courses that reached 84,528 workers, accounting for 1,095,405 contact hours of health and safety training.
• DOE nuclear weapons cleanup training—Seven awardees presented 896 courses to 14,097 workers, which resulted in 191,126 hours of safety and health training.
• MWT—Ten awardees presented 160 courses to 241 students, representing 215,767 contact hours. This program included over 27 different training courses or subjects. Since 1995, the MWT program has successfully trained 919 urban young adults throughout the United States in preparing them for career-path jobs related to environmental cleanup.

Mine Safety and Health Administration Education and Training

Mining in its many forms had historically been one of the most dangerous of all occupations. The Federal Mine Safety and Health Act of 1977 (the Mine Act) established the MSHA to enforce compliance with mandatory safety and health standards as a means of eliminating fatal accidents, reducing the frequency and severity of nonfatal accidents, and promoting improved safety and health conditions in the nation's mines. Just as OSHA carries out the mandates of the Occupational Safety and Health Act of 1970 for most industries other than mining, MSHA carries out the mandates of the Mine Act at all mining and mineral processing operations in the United States, regardless of size, number of employees, the commodity mined, or the method of extraction. Although MSHA's program rests upon its congressional mandate to enforce the Mine Act firmly and fairly, the agency has long held that enforcement alone cannot solve all safety and health problems and has strongly emphasized the education and training of miners and managers in mine safety and health requirements.

MSHA requires that each U.S. mine operator have an approved plan for miner training. This plan must include

• 40 hours of basic safety and health training for new miners who have no underground mining experience, before they begin work underground;
• 24 hours of basic safety and health training for new miners who

have no surface mining experience, before they begin work at surface mining operations; and

• 8 hours of refresher safety and health training for all miners each year and safety-related task training for miners assigned to new jobs.

The National Mine Health and Safety Academy

MSHA's National Mine Health and Safety Academy at Beckley, West Virginia, is the world's largest educational institution devoted solely to safety and health in mining. The academy serves as the central training facility for federal mine inspectors and mine safety professionals from other government agencies, the mining industry, and labor. The academy's physical plant has classrooms, a simulated mine and laboratories that can accommodate up to 600 students, a large auditorium, cafeteria, gymnasium, and a residence hall with dormitory space for more than 300 people.

Courses are offered on safety and inspection procedures, accident prevention, investigations, industrial hygiene, mine emergency procedures, technology, management techniques, and other topics. Courses range in duration from 2 hours to 8 days. From fiscal year 1994 through fiscal year 1998 an average of approximately 10,000 students per year took academy courses, either at the academy or at their work site. About 4,000 of these typically were MSHA and other government personnel, and about 6,000 were industry or union representatives (Thomas MacLeod, MSHA, personal communication, July 23, 1999).

Besides providing classroom instruction, the academy staff produces videotapes, films, publications, and a wide variety of technical materials. The academy also provides field training and serves as a technical resource to help meet the mining community's instructional needs.

Educational Field Services

In 1998, MSHA created the Educational Field Services (EFS) program to optimize the administration's resources for improving health and safety training for the mining industry. EFS training specialists visit mine sites and work closely with mine management, miners, and mine instructors to develop training methods to improve safety and health. These specialists then coordinate agency resources to best meet each mine's individual needs.

In addition to mine visits, EFS training specialists work with mining associations, safety organizations, labor unions, and educational institutions to establish partnerships and network resources.

Training Materials

MSHA makes available many training publications, manuals, courses, films and videotapes, and other materials not only through the Mine Academy but also through the MSHA's district offices. Some of the products available include

- small mine operator safety kits that assist operators of small underground coal mines in controlling workplace hazards generally found at mines that employ 50 or fewer workers;
- training materials on safe operation of powered haulage targeted to industry personnel who operate haulage trucks or other types of mobile surface mining equipment;
- training materials on health issues, including silicosis, diesel exhaust gases and particulate matter, pneumoconiosis (black lung), and other respirable dust problems;
- best practices techniques developed cooperatively by labor, management, and government groups;
- comprehensive training modules and a videotape on accepted job safety analysis methods and step-by-step miner task training procedures for use by supervisors and forepersons in diverse types of mining;
- course materials that review basic ventilation principles and practices for underground coal mines;
- a training program that features slides that illustrate fatal accidents in mining in recent years; and
- a monthly bulletin that features articles on topical health and safety issues.

Degree Programs in Mining Engineering

Although in many respects MSHA worker training could be a model worth emulating in other industries, it should be clear from the previous paragraphs that MSHA does not provide any support for university-based educational programs that lead to graduate or undergraduate degrees in mine safety or health. In fact, it has become apparent that university programs in all aspects of mining are in crisis. Mining engineers are educated to design and often manage all aspects of mining enterprises, including the health and safety of mine workers.

Eighteen years ago, 27 schools had ABET-accredited 4-year mining engineering programs, and the programs were viable in university academic environments. Today, 15 schools have ABET-accredited 4-year mining engineering programs, and the programs are hard pressed to continue in very cost-conscious university academic environments. Only one program has an undergraduate enrollment of more than 100 (Virginia Poly-technic Institute), and it has a small graduate enrollment, reflective of the recently changed loss of opportunities for mining-specific research funding. The average undergraduate enrollment among the 15 accredited programs was 51 in 1997, but 9 of these programs had less than 50 students. Clearly, market demand for mining engineering students, reflective of intense global competition, has diminished. There also appears to be little demand for students with higher mining engineering degrees in industry, and traditional research funding avenues have vanished, thereby reducing graduate student enrollments.

As a result of this global competition the survival of all but six programs is seriously threatened. The other six programs are also falling under intense scrutiny because of low enrollments and seriously diminished research funding. This scrutiny by university administrators has resulted in reductions of mining engineering faculty to levels just sufficient to maintain accreditation. Further down the road, it is likely that the weaker programs will not survive as senior faculty members retire and are not replaced. Senior professors to the mining schools are older and will be retiring in the next few years, and it has already been difficult to attract new faculty at mining schools that have been permitted to do so (e.g., the University of Utah mining program has twice advertised unsuccessfully for a new faculty member with coal mining experience).

MSHA, labor unions (e.g., the United Mine Workers of America), mine operator associations (e.g., the National Mining Association and the Bituminous Coal Operators of America), and NIOSH are all concerned that skilled mining professionals who will manage mines to protect the health and safety of miners will not be available when needed. The Mine Safety and Health Research Advisory Committee, at its June 10, 1999 meeting, asked NIOSH to examine the weakened state of the nation's mining schools and their ability to produce qualified personnel in health and safety matters and make recommendations on models of research and education funding that would address this problem. Such an examination of a single industry is beyond the scope of the present report, but the committee recognizes the reported difficulties of the mining engineering field as an instance of the more general problem of the occupational safety field in attracting and retaining doctoral-level scientist-educators to train future generations of safety practitioners.

Industry Training Programs

Corporate America provides nearly 2 billion hours of training to approximately 60 million employees, at a cost of nearly $60 billion (*Training Magazine*, 1997). Only part of that training is health and safety related, but large companies generally recognize the value of maintaining professional skills that support the well-being of their employees and workplaces. In a nonrandom survey of 10 health and safety-oriented Fortune 500 companies (Victor Toy, International Business Machines Corporation, personal communication, September 1999), skills development was considered an important priority for health and safety professionals. Although it is not representative of the private sector as a whole, the results of this informal telephone survey illustrate the range of training activities that are undertaken in some large, multinational corporations.

The companies surveyed included companies in the petrochemical, high technology, aerospace, chemical, and pharmaceutical industries that ranged in size from 14,000 to more than 300,000 employees and that employ three to several hundred health and safety personnel. Programs for training and skills development differed between companies. For example, at one end of the spectrum one company focused training in a single specific area, and on the other end of the spectrum another company managed skills more comprehensively on the basis of the identification of strategic requirements to support projected business growth and development on a worldwide basis. All companies surveyed included skills training as an aspect of individual, annual performance reviews, which made it possible to require assignments and educational opportunities for the continued growth of the OHS professional.

Some company training programs included a mapping of skills against the needs of the specific businesses in the corporation. For instance, International Business Machines Corporation (IBM) developed a rigorous process that requires OSH professionals (i.e., industrial hygienists, safety engineers, doctors, and nurses) to attain specified levels of predetermined competencies on the basis of job descriptions aligned to meet a broadened array of employee well-being needs for a global business. In this approach to training and skills management, the employee establishes an annual training and education plan that is validated to meet the predefined skills milestones, and then the plan becomes funded, is executed and is monitored as the individual's development plan.

The intensity of OSH education and training programs varied widely between and within corporations. Business performance, for example, could cause short-term program changes. Some companies hire people with the skills they need rather than invest heavily in continuing education. Every company surveyed provided at least 1 week of continuing education for each professional per year. Some provided as much as 4 weeks of training in a calendar year. Most also provided additional training and education on an as-needed basis to support the work of the OSH professional. Technical training is generally obtained through external resources locally or nationally, with most being provided by associations like the National Safety Council, the American Society of Safety Engineers, the American College of Occupational and Environmental Medicine, the American Association of Occupational Health Nurses, and the American Industrial Hygiene Association. Management and business training for OSH leaders and managers, on the other hand, was generally provided through internal corporate organizations. Tuition reimbursement for academic course work and degree programs surveyed was available in nearly all companies.

Corporations with sizable OSH staffs may also develop and conduct internal conferences, training programs, and courses. The format varies from 1-day sessions on a single subject to a full week of training on-site or at a central location. Session topics are issues most relevant to current business operations, although basic training in OSH fundamentals is sometimes provided. One company in the petrochemical industry had a unique program to ensure the development of core skills in its OSH staff. Its "campus hire" program, which follows a structured health and safety curriculum, seeks to ensure that a professional acquires core OSH competencies during the first 2 years before a permanent job assignment is given.

Health and safety-oriented multinational corporations recognize well the limited availability of educational opportunities and OSH skills outside of mature industrialized countries. In these instances, companies often use training programs that develop the necessary skills of local employees through a combination of class and field work. A common format involves senior OSH staff members providing individualized training over a period of weeks to months with on-going mentoring support. In some cases, training plans must be approved by host countries looking to retain employees with OSH skills in the region.

Corporations promote skills development by supporting with time or funding the acquisition or maintenance of a professional certification(s) and the provision of internship experiences or rotations. Nine of the 10 companies interviewed considered professional certification in an employee's development plan. In two cases, certification was necessary to attain a certain level within a job family. One company also rewarded this achievement through compensation. Internships and rotations are common in industry and provide students and physicians in training with private-sector experiences.

Union- and Labor-Sponsored Training Programs

Unions have long been involved in safety and health training. These efforts were spurred on by the OSHA New Directions program in the late 1970s, in which a few million dollars in grants were given to unions and trade associations to develop training materials for workers and to provide training. A huge amount of training materials and training was generated from those initial grants. Many safety and health programs were started at unions and trade associations as a consequence of this seed money. That program ended in the early 1980s and was subsequently replaced by smaller more targeted grant programs (e.g., programs targeted to certain hazards or hazardous industries) such as the Susan Harwood Training Grants Program in the 1990s.

Some unions, such as many unions for construction workers, offer all new workers safety training as part of their apprenticeship programs, such as the building trades standardized 10-hour OSHA training curriculum called Smart Mark. Several unions have set up joint safety and health funds, financed by employer contributions of so many cents per hour. Examples of such funds are the United Automobile Workers joint funds with Ford and General Motors, the Laborers' Health & Safety Fund of North America, and the Laborers-AGC (associated general contractors) Training Fund. These funds develop safety programs that are then used to train members. The Laborers-AGC Training Fund, for example, has more than 70 training centers in the United States and Canada. These centers provide both skills training and safety training to more than 800,000 members. Much of that training is provided to workers before their employment at a specific work site, avoiding the otherwise high risk of injury in the first few days on the job as well as ensuring that an increasingly transient workforce receives this important training regardless of the nature of the employment contract and employer values. The safety and health departments of many international unions not only provide training, but also facilitate joint health and safety committees, provide technical information and conduct inspections, and advocate for stronger state and federal safety and health legislation.

Many programs that offer safety and health training of workers are sponsored by nonprofit organizations, (such as Committees of Occupational Safety and Health [COSH] that consist of coalitions of local unions in an area or city working together on occupational safety and health issues) and labor education programs at universities such as the Labor Occupational Health Program (LOHP) at the University of California at Berkeley. LOHP has, for example, been in the forefront of efforts to train minority workers, non-English speaking workers, and young workers. School-to-Work programs have been a successful innovation. Much of the training by COSH groups and labor education programs is funded through state and federal grants.

FUTURE NEEDS IN WORKER AND EMPLOYER
SAFETY AND HEALTH TRAINING

Although this brief review provides only a hint of the number and variety of OSH training courses and programs in use or available today, it is clear from the review and, for example, the October 1999 National Conference on Workplace Safety and Health Training, that the volume of training being conducted is substantial. However, as noted above, with rare exceptions, little is known about the quality of the training that is actually conducted (does it consist of an actual class, a videotape, or just a handout?). Similarly, there is little information on the effects of such training; for example, did the incidence of workplace injuries or illness decrease after training, and what distinguishes successful training from unsuccessful training? Such questions are seldom even asked. Cohen and Colligan (1998) found 80 reports in the published literature, spanning the years 1980 to 1996, in which training was explicitly evaluated as an intervention effort. They concluded that training had universally shown merit in increasing worker knowledge about job hazards and effecting safer work practices, but because training was often coupled with other interventions, a connection to decreased numbers of injuries, a decrease in the amount of time lost from work, or decreased medical costs was never clearly established. The authors also emphasized that because few of the studies reviewed actually manipulated variables like class size, length or frequency of training, trainer qualifications, mode of training, and management involvement, it was impossible to say what factors produce the greatest impact.

Training and education of workers has not traditionally been considered a prime responsibility of most OSH professionals. In fact, most graduates of OSH programs are ill prepared for this assignment. While they have the technical knowledge, many graduates lack skills in adult education, training, and program evaluation. The net result is very uneven training quality. As noted in this chapter, several OSHA, NIOSH, and NIEHS programs have attempted to fill this gap; but these programs have often been limited to selected populations (e.g., hazardous waste workers, emergency responders, etc.) by their enabling legislation. In addition, the inability of these programs to reach a significant portion of the nonunion workforce is cause of concern as well as impetus for seeking innovative approaches, perhaps community based, to reach these workers.

The committee urges NIOSH to join with OSHA, NIEHS, unions, industries, and employers to systematically evaluate the efficacy of OSHA and other worker training programs and better define minimum training requirements. Using the successful collaboration that has characterized the National Occupational Research Agenda (NORA) as a prototype, these agencies should undertake a major effort to: (1) explore how lessons learned from these programs can be used to enhance other worker training efforts, and (2) broaden the scope of worker populations which can benefit from these substantial expenditures of funds. Demonstration project grants should be provided as incentives to develop model training programs for OSH educators and trainers in specific employment sectors.

Earlier sections of this report have noted that OSH professionals are primarily employed by medium-sized to large employers and therefore directly involved with less then half of all U.S. employees. The committee's analysis of anticipated changes in the nature of work in the United States led it to suggest that small, service-sector businesses would increasingly dominate the employment market and that all sectors of the economy would be characterized by decentralization, flexibility, non-standard work arrangements, and a highly diverse and transient work-force, making it likely that an even smaller proportion of employees will count OSH professionals among their fellow employees. Day-to-day responsibility for worker safety and health will increasingly fall to managers who have little if any formal education or training in OSH and who may have numerous other responsibilities as well. NIOSH and OSHA should collaborate to develop a program of training for these individuals. The committee believes the two agencies need to reconsider their historical division of training responsibilities (OSH professionals by NIOSH, workers by OSHA) in the light of this trend. The OSHAct of 1970 is far less specific on this division than the agencies' policies might imply (see Chapter 1). The committee again recommends large-scale demonstration projects that target small- and medium-sized employers and encourages the use of new learning technologies, the development of a recommended set of basic competencies, and the creation or recognition of a new category of OSH personnel, the occupational safety and health manager. The partnering process so successfully used by NORA could be usefully employed in this endeavor as well.

Incentives may still be necessary to induce small businesses in industries not currently covered by specific OSHA standards to invest in either high-quality worker training or education of an OSH manager for their work sites. This is an important consideration, given that it is just these sorts of employers that will be the primary source of new jobs in the coming decade. Clarification of existing OSHA training mandates to include essential elements and measures of efficacy may be one answer to the training quality issue, and OSHA's draft proposed Safety and Health Program Rule unequivocally fixes responsibility for hazard identification and control and for worker health and safety training on the employer and management. Further inducement depends upon inculcation of a culture of safety and health in the general public. Accomplishing this is a large, long-term, multifaceted project that will require leadership from the U.S. Department of Health and Human Services and the U.S. Department of Labor and that will involve the mass media, the Internet, K-12 education, and other communications strategies targeted to youth, parents, and workers as well as employers.

8

Alternatives to Traditional Classrooms

ABSTRACT. The Internet is one of a number of alternatives to full-time attendance in traditional education programs for midcareer professionals seeking information about occupational safety and health (OSH) for immediate use or in pursuit of an advanced degree. Distance education and distance training are examples of means of instructional delivery that afford the learner the opportunity to engage in learning experiences away from the traditional classroom. They are planned and structured means of learning that use electronic technology-based media including audio, print, video, and the Internet, alone or in combination. Limited but impressive data on the popularity and effectiveness of distance education in preparing physicians for occupational medicine board certification examinations point to its potential as a means of facilitating education and certification of the many practicing OSH personnel without formal specialty training in the area. The committee concludes that, although traditional approaches remain indispensable for some types of instruction, the National Institute for Occupational Safety and Health (NIOSH) should develop incentives to promote the use of distance education and other nontraditional approaches to OSH education and training. An essential part of these innovative programs should be thorough evaluation of both the program content and the performance of their graduates in relation to the performance of graduates of traditional programs in job placement and on measures like certification examinations.

As noted in Chapters 2 and 7, doctors and nurses often discover occupational medicine and occupational health nursing only after graduation and experience in another field. Occupational safety professionals may also discover the field only after considerable experience in another job, very often in a high-risk industry. In all of these cases, return to school for formal education as a full-time student is generally not a feasible option. Innovations like "executive" master's degree programs that pack course work into a series of all-day weekend meetings have been one response. The On-the-Job Off Campus Program at the University of Michigan is one example of this approach to training working professionals. Students attend classes at the University for four days once per month for 22 months. Other concentrated programs, like the North Carolina Education and Research Center's Winter and Summer Institutes, offer industrial hygiene technician and safety technician certificates for completion of six 1-week or half-week courses over a 3-year period. For people who are responsible for health and safety as an ancillary duty or who are contemplating a career change this may be a more appealing alternative than pursuing a degree or CIH/CSP certification.

The Internet offers another approach to the information needs of OSH personnel, and employers as well as those of workers. Both NIOSH and the Occupational Safety and Health Administration (OSHA) have been working hard to take advantage of the Internet's vast potential for providing ready access to information, data, and tools for improving occupational safety and health. The OSHA website (www.osha.gov) has thousands of pages of regulations, publications, and other documents online, including the latest versions and updates of all health and safety standards. A huge advantage of accessing this information via the Internet is that ever more powerful and user-friendly search engines can be used to rapidly pull out all the documents of interest. An "expert adviser" is interactive diagnostic software that asks the user a series of questions and follow-up questions to determine whether, how, and which specific parts of an OSHA standard apply to the user's activities. On the basis of the answers that the user gives, the adviser determines what information the user needs to know about the standard's application to the user's activities. The NIOSH website (www.cdc.gov/niosh) is also packed with useful information for the busy OSH professional. Searchable publications and databases such as the *Pocket Guide to Chemical Hazards* (National Institute of Occupational Safety and Health, 1997b) and the *Manual of Analytical Methods* (Cassinelli and O'Connor, 1994) are supplemented by Health Hazard Evaluations, notices of upcoming meetings and conferences, descriptions of research and training grant programs, and links to a myriad of other websites. Listservers such as the Duke University Occupational Environmental Medicine Mail List Health (http://gilligan.mc.duke.edu/oem/occ-env-.htm) can be used to query hundreds of other subscribers on their experience with any aspect of OSH.

Another alternative to traditional classrooms is distance learning or distance education. Distance education and distance learning modalities offer flexibility for reaching individuals beyond the traditional classroom and campus or large corporate training programs, and many organizations, public and private, have already begun to incorporate computer technology into worker training. Distance learning is used at many education levels, ranging from workplace training for specific preventive measures related to a hazardous substance to certificate, associate, baccalaureate, and graduate degree programs and continuing professional education. Technology-based education is expanding rapidly, with the expansion spurred by both the job needs and higher education institutions' efforts to meet the needs of a changing workforce and student population. Distance learning programs in higher education programs initially focused on part-time students and nontraditional students located at a distance from the campus. These programs are now expanding and are being integrated into more traditional campus-based programs (National Center for Education Statistics, 1997).

Many state education systems have active distance education programs, including the states of Maine, Colorado, and Kansas. The governors of 15 states are developing a "virtual university" that will have no physical campus but will use computers and interactive video to provide instruction (National Center for Education Statistics, 1997). The Western Interstate Commission for Higher Education has developed a cooperative arrangement for educational telecommunications that has 12 participating institutions, including Pennsylvania State University, Ohio State University, and the Universities of Iowa, Minnesota, Wisconsin, and Illinois (National Center for Educational Statistics, 1997). The National Institute of Environmental Health Sciences (NIEHS) held a workshop in April 1999 to assess the potential use of "advanced training technologies" for hazardous substance training. The ensuing report (National Institute of Environmental Health Sciences, 1999) included a comprehensive analysis of existing and emerging technologies and concluded that NIEHS's call for fiscal year 2000 hazardous materials training grants should include encouragement of applications for programs that pilot the use of advanced training technologies.

With the need for expanded training of workers in OSH, many of whom work in small establishments, and the need for professionals at every level, increased use of distance learning can help fill many of the currently unmet education and training needs. Distance learning is explored here as a potential means of facilitating lifelong learning by OSH professionals and more specifically as an alternative to traditional graduate training in occupational medicine and occupational health nursing, two professions with large numbers of practitioners without formal specialty training in the area.

DISTANCE EDUCATION MODALITIES

A number of different technologies, alone or in combination with other modalities, can provide distance education. The oldest form of distance learning is the correspondence course. The following are current technology-based delivery media:

- *Virtual training* is used to enhance and support real-world training. It replicates real-life situations by appearing lifelike and three-dimensional, but it trains the learner without risking potential dangers. A well-known example is the aircraft simulator, which is routinely used to train pilots in emergency procedures. High-voltage training simulators, mining simulators, vehicle simulators, and medical laboratory simulators are used to train individuals safely, allowing for errors, eliminating casualties and saving revenue by avoiding destruction of property.
- *Internet video conferencing* allows the transmission of digital voice and video through the Internet. The strengths of this medium include its convenience, high degree of accessibility, economy, and interactivity. Participants have the ability to receive video conferencing broadcasts at their desktop computers and also have the ability to communicate synchronously. It does require student participation at specific times, however.
- *CD ROM* is an advanced upgrade of the audio compact disc (CD). It stores computer-readable programs, images, and digital audio data. CD ROMs have large storage capacities, are fast, and if properly handled, will last indefinitely. Students can use them at their own convenience. Their shortcomings include an inability to erase, modify, and update the information on the disk.
- *Interactive videodiscs (IVD)* is the CD laser version of the earlier 16- and 35-mm training films. IVDs include features of CD technology such as long life and high data storage capacity. IVDs can store as much as 54,000 frames on each side of the disc and each frame may be viewed individually or in forward motion and reverse. They share the same strengths and weaknesses as CDs.
- *Teleconferencing* describes the interaction between instructors and learners. Teleconferencing includes audio, audiographics, video, and computer interaction. Audioconferencing connects participants through telephone calls and is relatively inexpensive. Audiographics adds a visual component (via fax or copier) to support instruction. Videoconferencing transmits voices and images through telephone lines and allows an immediate synchronous interaction between all participants. Computer conferencing connects through computer networks. Instructors and learners interact, primarily through electronic mail, and their interaction can be

synchronous or asynchronous. Like Internet videoconferencing, the methods demand student attention at specific times.

• *Computer managed instruction (CMI)* allows the learner to gradually work through training modules. On the basis of their performance and skill mastery, learners can move forward or backward through the training module. CMI instruction is self-paced for optimal understanding and does not require travel or absence from work (Stephens, 1999).

DISTANCE EDUCATION AND WORKFORCE RELATIONSHIPS: POTENTIAL BENEFITS AND COSTS

Distance education expands the ability to offer education toward a certificate or degree, continuing education, and work site education and to expand the pool of professionals and on-site employees and employers in OSH. The different techniques provide individuals with the flexibility, depending on the modality, to learn at the work site, at home, or in local community facilities at specially established network sites. Distance learning modalities have and can be used in degree programs like the Master's in Public Health offered by institutions such as the Universities of Wisconsin and Kansas or to provide specific instruction like that being developed by NIEHS in the handling of hazardous materials and the United Auto Workers' demonstration of the online Right to Know computer network or refresher and continuing education training (National Institute of Environmental Health Sciences, 1999).

The OSH workforce, on average, is technically-adept, well-educated, and has a low unemployment rate. In many respects, OSH professionals are typical, or even ideal, "adult learners." The adult learner views education as a mechanism that can be used to reach specific career goals that can include specific skill enhancement, continuing education, and degree and advanced degree education. Most adult learners are employed full- or part-time, have family responsibilities, and cannot readily participate in traditional classroom-based education. With the flexibility of the training capabilities and sites of training, distance education can significantly build the skills of employees, managers, and students at various academic levels. Health professionals can also participate in continuing education in almost any environment (Stephens, 1999).

Distance education requires careful planning and matching of the education and student requirements with the different modalities. For example, if part of the training requires hands-on exercises or preceptorships with a mentor, the distance learning can fulfill only part of the teaching needs and augmentation will be needed to either bring the students to mentors or mentors to the students. Curricula need to be carefully designed for the particular modality, evaluated, and modified. With methods like videoconferencing and teleconferencing, faculty usually require some training to use the medium effectively.

Distance leaning alone cannot always substitute for traditional classroom or onsite hands-on experiences. Many distance learning programs are augmented by periodic meetings of students and faculty mentors to encourage interpersonal interaction and dialogue. For certain skill training, including industrial hygiene analytic techniques and patient care in medicine and nursing, hands-on work in the presence of faculty or preceptors is still required. For some students and trainees, particularly younger students, interaction with other students and faculty on the campus or at the work site remains important.

Determination of the costs of distance learning and who should bear the costs is often difficult and confusing, primarily because of the complexity of the technology. It is necessary to distinguish among technologies with costs that vary widely between one-way and two-way technologies. With one-way technologies, consideration needs to be given to the additional costs of tutorial support, in contrast to two-way technologies, for which the tutorial costs are factored into the base cost. Costs will vary with the student numbers and with the need to develop new course materials or adapt or adopt existing ones.

The initial capital costs of the technologies also vary. For example, if computers are available to all students the costs will be quite different than if computers need to be purchased or upgraded. There are substantial costs of installing telecommunications lines and video equipment. Certain costs are one-time capital investments or fixed costs like developing the education materials and courses. Other costs like faculty time for teaching and mentoring, grading examinations and papers, and evaluating the content and outcomes of the course, telephone costs, and costs maintenance of equipment are ongoing, and again the unit cost will vary with the number of students and whether some costs are no longer incurred with the new education or training technique. For example, if distance learning reduces the number of faculty previously needed to teach students in different locations or times of day, these savings will offset some of the costs of the new teaching modality.

Cost efficiency is related to program goals, objectives, and outcomes and how they are matched with the most appropriate technology. These technologies, as will be seen in some of the examples that follow, take advantage of economies of scale. Without careful planning, however, costs can become inefficient. Because distance education can be developed to meet very specific criteria and "one size will not and does not fit all" there is no standard pricing. Choices of media include CD-ROM, virtual reality, teleconferencing, World Wide Web-based courses, and satellite broadcast, with the initial and ongoing costs varying with the media selected.

Compared to traditional face-to-face modes of instruction in the classroom, distance education has potential cost-savings advantages including the indirect costs incurred by students in travel, travel time, lost wages, and so forth. Although there may be high initial costs in establishing distance learning infrastructures, the long-range cost savings of open learning (no qualifying academic requirements) may be attractive alternatives for companies. For example, a traditional face-to-face training course with 1,000 participants that met for 2 days at a conference center can cost considerably more than the same conference deployed electronically and with less productivity loss.

Costs can also be affected by such things as shared networks or courses, where multiple institutions collaborate on joint educational and training activities and either use existing sites for the distance learning or establish decentralized education networks as in the case of the Kansas universities discussed later or the virtual universities referred to earlier. Infrastructure costs are substantially reduced with broad-based cross-institutional collaboration (Stephens, 1999).

CURRENT EXAMPLES

Described below are examples of education and training programs delivered by distance education methods. These programs are using a variety of technologies shaped by the audiences they are trying to reach and other factors.

The University of Kansas

The School of Nursing at the University of Kansas (KU) has been a pioneer in developing distance learning master's-level programs. There are two types of programs: one that is entirely KU based and another that is collaboration with other nursing programs across the state.

Using KU's *Virtual Classroom,* the School of Nursing is bringing the classroom to students' homes. The term *virtual classroom* describes an Internet website on the KU computer network. This website houses courses for Schools of Nursing, Medicine, Allied Health, Pharmacy, Graduate Studies, and Continuing Education. A virtual classroom does not require a student to have a physical presence in a traditional classroom. Students can then choose when, where, and how to learn within a prescribed set of criteria. The School of Nursing has the following programs available through the virtual classroom:

• A Master of Science in nursing with a major in family nurse practitioner or adult/geriatric nurse practitioner for Bachelor of Science in Nursing-prepared nurses who have already completed certificate programs in these fields. The courses are offered entirely on the World Wide Web.

• A Master of Science in nursing with a major in family nurse practitioner at sites in Kansas City and Garden City, Kansas. This program is geared for nurses not currently prepared as nurse practitioners. Because this program combines World Wide Web-based courses and interactive television, some classroom work is required in one of the cities mentioned above.

The School of Nursing has also developed master's level nurse practitioner programs with other nursing schools in the state including Pittsburg State University, Fort Hays State University, and Wichita State University. There is a single faculty with faculty members based at the different institutions and courses presented through teleconferencing, mainly from the KU campus. Students take courses through the televideo at their home campus, and the clinical clerkship requirements are met locally.

The same philosophy underpinned the joint Master of Public Health program that is a collaborative program between the University of Kansas Medical School and the College of Health Professions at Wichita State University. An off-site campus for the convenience of state personnel was established in Topeka. Faculty from both universities teach courses on all campuses by televideo, with KU specializing in epidemiology and biostatistics and Wichita State specializing in social and behavioral science and administration. The two universitites are planning an expansion of the joint program to provide access to certificate-level courses statewide for public health personnel who have no formal training using distance learning.

The KU Medical School has developed network education sites in six locations in Kansas. These sites are equipped with computers and televideo capacity and compressed video. Combined with the efforts of the Area Health Education Centers, these sites are designed to provide distance education for numerous health professionals who either are taking clinical preceptorships in the area or want to enroll in continuing education or in degree and certificate programs. A master's-level occupational therapy program in collaboration with Pittsburg State University is currently offered at one site. (For more information, see the program description at http://www2.kumc.edu/son/prospective.htm.)

Medical College of Wisconsin

The Medical College of Wisconsin (MCW) provides nationwide access to a distance education Master of Public Health program for physicians. The mission of the program is to provide didactic courses that may be adjunct to clinical training programs and that may partially satisfy prerequisites for board certification in preventive medical specialties including OSH. Occupational health students were first accepted in 1986, with the Master of Public Health program accredited in 1991. The entire curriculum is based on distance learning using computer technology. Each course has a set of goals and objectives, an introduction to each module, required and supplemental reading lists, activity assignments, computer quizzes, and report forms for activity assignments. A final examination is administered by a local (to the student) proctor at one of the 600 local institutions with which MCW has contracted. There is an 8-hour on-campus orientation on the mechanics of the program; the only other time that students are required on the campus is for the graduation ceremony. The completion of 10 courses is required for the degree. The courses are intended to be 4 months in length, and are generally taken one at a time. The program is self-paced, and although at least one student has graduated in a year, the average time to graduation is 4.5 years. As of June 1999 there were 347 graduates of the program, and 250 students were currently taking a course. As noted in Chapter 7, MCW graduates have an exceptionally high rate of passing the occupational medicine certification examination ("boards") of the American College of Preventive Medicine (For more information, see http://instruct.mcw.edu/prevmed/).

University of North Carolina at Chapel Hill

The University of North Carolina School of Public Health at Chapel Hill has a Master of Public Health Leadership degree program that aims to provide access to higher education for professionals and health-related organizations via the Internet and Internet-based conferencing. The project uses cohorts of students to eliminate the lone student approach. Most courses are taught by using a combination of videoconferencing and World Wide Web-based instruction. Groups of learners gather each Thursday evening at seven sites in North Carolina for two-way interactive videoconferences with professors and students at other sites. Professors give lectures and lead group discussions within and across the sites. Students also receive readings and lectures over the web and interact with the professors and other students by electronic mail and computer-based discussion forums. The School of Public Health also offers a master's degree (MPH) in public health nursing with a concentration in occupational health nursing. Beginning in August 1999, the program has included a distance learning option that is Internet-based and requires only 2.5 weeks of campus coursework each year (The program can be completed in 2 years, but students are allowed up to 5 years). (For more information, see http://cdlhc.sph.unc.edu/phl).

The National Technological University

The National Technological University (NTU) is a consortium of institutions that share courses and that offer a large variety of degrees. Courses are delivered directly to the work site (coupled with the ability for recording of broadcasts for student viewing at home). NTU students are employees of organizations that have installed equipment that receives signals from the NTU satellite. The downlink equipment allows employees to access NTU's course offerings via satellite.

By means of instructional television, engineers, scientists, and managers at their job sites can tune in to technical and managerial courses offered by top faculty and experts at the nation's leading engineering schools and other organizations selected for their expertise. More than 500 academic courses are offered. NTU offers 14 master's degree programs designed specifically for technical professionals (see http://www.ntu.edu).

Other Programs

There are numerous additional examples such as the associate degree program in occupational health and safety at Trinidad State Junior College that has recruited workers who have been injured on the job and return to their industries as OSH workers. The U.S. Army Corps of Engineers has developed an 8-hour World Wide Web-based hazardous waste operations and emergency response refresher training program. Tulane University School of Public Health has announced plans for a new Internet-based Masters in Public Health degree program in Occupational Health that is designed for physicians, nurses and other health professionals who work in occupational health programs or clinics. The two-year, part-time program will provide the academic year required for board certification in preventive medicine/occupational medicine. The course content in this real-time interactive program will be the same as the on-campus program, but students will attend class over the Internet by logging onto the course website and receiving an audio broadcast from the instructor, delivered through a combination of pre-recorded lectures and live instructor-led class sessions.

EDUCATION OUTLOOK FOR OCCUPATIONAL
SAFETY AND HEALTH

Distance learning technologies have the capability of meeting the educational needs at all levels for the field of OSH. From short training modules at the workplace or home to full graduate degree programs, the distance learning technologies can be used alone or in tandem. The advantages of distance learning include the flexibility of the sites of learning, the ability to draw on expertise from multiple institutions to tailor the education content to the need, the ability to use multiple modalities to reinforce traditional classroom-type activities, economies of scale, and replicability. Disadvantages include infrastructure costs for techniques such as teleconferencing, absence of hands-on skill instruction, potential isolation of students, faculty resistance and the need to train faculty to use distance learning modalities, lack of capability for informal "corridor consultation," and potential high direct costs depending on the technology used and the number of students enrolled.

CONCLUSION

With the dramatic changes that have occurred and that will occur in the workplace and in the composition of the workforce, flexible new methods of training workers and professionals in occupational health and safety need to be considered for all environments. Distance education techniques can be used not only to transmit discrete modules on specific toxic substances but also to provide access to degree programs and ongoing continuing education for busy OSH professionals. The learning can take place in the home, at the work site, at local community sites, or on college and university campuses.

The committee concludes that although traditional approaches remain indispensable for some types of instruction, NIOSH should develop incentives to promote the use of distance education and other non-traditional approaches to OSH education and training. An indispensable part of these innovative programs should be the thorough evaluation of both the program content and the performance of their graduates in relation to the performance of graduates of traditional programs in such areas as credentialing examinations and job placement.

9

Summary of Findings and Recommendations

The major trends that the committee has identified in the workplace and the workforce, most of which arise from the movement away from a long-term employee-employer relationship at a fixed work site, are elements of larger societal changes related to the way social services are delivered in modern industrial economies. Destabilization of employment relationships has led to increasing movement toward making the individual even more responsible for his or her own financial welfare over a lifetime. The notion of altering the social security system to allow individuals to manage their own funds in individual retirement account-like schemes is a recent example. The traditional provision of health insurance and a retirement program through the employer is also becoming less prevalent. Moreover, public support to strengthen mechanisms to provide these services through governmental programs appears to be declining as well. Churches, labor unions, and fraternal organizations cannot fulfil the role of social services conduit on a broad scale. Indeed, labor unions themselves are struggling to survive in sectors of the economy where employment is increasingly transient. The end result is that the evolving circumstances of employment are increasingly placing responsibility for occupational illness and injury prevention on the individual worker. The committee does not believe that this trend is in the best interest of U.S. workers and has thus devoted much of its analysis and recommendations to actions that will counter or mitigate this development instead of a more straightforward calculation of the occupational safety and health (OSH) personnel likely to be needed by the large, fixed-site, stable-workforce industries that have traditionally hired them.

Enhanced and innovative approaches to OSH and education are needed at the graduate level for health and safety managers, for industrial sector-specific training for those doing worker training, and for the general public with an emphasis on reaching children, parents, and the workforce. The committee makes the following specific recommendations, numbered for ease of reference and not as an indication of priority.

CURRENT OSH WORKFORCE AND TRAINING

The current supply of OSH professionals, though diverse in knowledge and experience, generally meets the demands of large and some medium sized workplaces. However, the burden of largely preventable occupational diseases and injuries and the lack of adequate OSH services in most small and many medium-sized workplaces indicate a need for more OSH professionals at all levels. The committee also finds that OSH education and training needs to place more emphasis on injury prevention and that current OSH professionals need easier access to more comprehensive and alternative learning experiences.

Recommendations

To address the critical need to mitigate the enormous and continuing impacts of acute and chronic injuries on worker function, health, and well-being, to develop new leaders in this neglected field, and to strengthen research and training in it at all levels:

Recommendation 1: Add a new training initiative focused on prevention of occupational injuries.

NIOSH should develop a new training initiative focused on the prevention of occupational injuries, with special attention to the development of graduate-level faculty to teach and conduct research in this area. Possible approaches would include regional Occupational Injury Research, Prevention, and Control Centers as an entirely new program or by modification of the existing NIOSH training programs or collaboration with the Centers for Disease Control and Prevention's National Center for Injury Prevention and Control.

To enhance needed multidisciplinary research in injury prevention and in occupational safety and health in general:

Recommendation 2: Extend existing training programs to support of individual Ph.D. candidates.

NIOSH should extend existing training programs to support individual Ph.D. candidates whose research is deemed of importance to the prevention and treatment of occupational injuries and illnesses, independent of academic department or program. Restricting support to students in Education and Research Centers or Training Project Grants–affiliated departments or disciplines deprives the OSH field of individuals who may have innovative responses to changing circumstances.

To address the lack of formal training among OSH professionals:

Recommendation 3: Encourage distance learning and other alternatives to traditional education and training programs.

NIOSH should encourage the use and evaluation of distance education and other nontraditional approaches to OSH education and training, especially as a means of facilitating education and certification of the many practicing OSH personnel without formal specialty training in the area.

Recommendation 4: Reexamine current pathways to certification in occupational medicine.

The American Board of Preventive Medicine should reexamine the current pathways to certification in occupational medicine. Specifically, it should consider

• extending eligibility for its existing equivalency pathway to include physicians who graduated after 1984 and
• developing a certificate of special competency in occupational medicine for physicians who are board certified in other specialties but who have completed some advanced training in occupational medicine.

FUTURE OSH WORKFORCE AND TRAINING

Expected changes in the workforce and in the nature and organization of work in the coming years will result in workplaces that will be quite different from the large fixed-site manufacturing plants in which OSH professionals have previously made their greatest contributions. The delivery of OSH services will become more complicated, and additional types of OSH personnel and different types of training than have been relied upon to date will be needed. Simply increasing the numbers or modifying the training of occupational health professionals will not be sufficient, since the primary difficulty will be to provide training to underserved workers and underserved workplaces. Traditional OSH programs must be supplemented by a new model that focuses on these workers and worksites.

Recommendations

To help ensure high-quality occupational safety and health programs for the full spectrum of American workers:

Recommendation 5: Solicit large-scale demonstration projects that target training in small and mid-sized workplaces.

NIOSH, in collaboration with OSHA, should fund and evaluate large-scale demonstration projects that target training in small and midsized workplaces. These innovative training programs should encourage the use of new learning technologies, should include a recommended core of competencies, and could lead to the creation of a new category of health and safety personnel—OSH managers.

Recommendation 6: Evaluate current worker training and establish minimum quality standards.

OSHA should join together with NIOSH, NIEHS, unions, industries, and employer associations to evaluate the efficacy of OSHA and other worker training programs and better define minimum training requirements.

Recommendation 7: Solicit demonstration projects to create model worker training programs for occupational safety and health trainers.

NIOSH, in collaboration with OSHA, should fund demonstration project grants that target specific employment sectors as an incentive to develop model training programs for another category of health and safety personnel—OSH trainers.

To address the challenges posed by the increasing diversity of the U.S. workforce:

Recommendation 8: Increase attention to special needs of older, female, and ethnic/cultural minority workers.

All aspiring OSH professionals must be made aware of ethnic and cultural differences that may affect implementation of OSH programs. In addition, because OSH programs are social as well as scientific endeavors, NIOSH, OSHA, NIEHS, other federal and state agencies, educational institutions, unions, employers, associations, and others engaged in the training of OSH personnel should foster and/or support efforts to provide a body of safety and health professionals and trainees that reflects age, gender, and ethnic/cultural background of the workforces that they serve. These organizations should also foster meaningful instruction on the aging process, the interaction of disabilities and chronic diseases with workplace demands, and communication skills to interact with minority workers and workers with low levels of literacy and those for whom English is a second language.

To prepare present and future OSH professionals to address continuing changes in the U.S. workforce, in the workplace, and in the organization of work itself as major determinants of workplace safety, health, and well-being:

Recommendation 9: Examine current accreditation criteria and standards.

Boards and other groups that accredit academic programs in the OSH professions, in conjunction with appropriate professional organizations, should carefully examine their current accreditation criteria and standards, paying special attention to the needs of students in the areas of behavioral health, work organization, communication (especially risk communication), management, team learning, workforce diversity, information systems, prevention interventions, healthcare delivery, and evaluation methods.

Recommendation 10: Broaden graduate training support to include behavioral health science programs.

NIOSH should broaden its graduate training support to include the behavioral health sciences (e.g., psychology, psychiatry, and social work) by developing and maintaining training programs in work organization and the prevention and treatment of physical and mental effects of work-related stress.

BOX 9-1 SUMMARY OF RECOMMENDATIONS

Current OSH Workforce and Training

1. Add a new training initiative focused on prevention of occupational injuries.
2. Extend existing training programs to support of individual Ph.D. candidates.
3. Encourage distance learning and other alternatives to traditional education and training programs.
4. Re-examine current pathways to certification in occupational medicine.

Future OSH Workforce and Training

5. Solicit large-scale demonstration projects that target training in small and mid-sized workplaces.
6. Evaluate current worker training and establish minimum quality standards.
7. Solicit demonstration projects to create model worker training programs for occupational safety and health trainers.
8. Increase attention to special needs of older, female, and ethnic/cultural minority workers.
9. Examine current accreditation criteria and standards.
10. Broaden graduate training support to include behavioral health science programs.

References

Accreditation Board for Engineering and Technology. 1999. Related Engineering Accreditation Commission Accredited Programs for 1999. [www document]. URL http://www.abet.org/accredited_programs/ RACwebsite.htm (accessed August 2, 1999).

Accreditation Council for Graduate Medical Education. 1999. Program Requirements for Residency Education in Preventive Medicine. [www document]. URL http://www.acgme.org/pm/pm.htm. (accessed May 28, 1999).

Altman, S., and D. Schactman. 1997. Should we worry about hospitals' high administrative costs? *New England Journal of Medicine* 336:798–799.

American Association of Industrial Nurses. 1976. *The Nurse in Industry*. New York: American Association of Industrial Nurses.

American Association of Occupational Health Nurses. 1999a. What is Occupational and Environmental Health Nursing? [www document]. URL http://aaohn.org/whatis. htm (accessed August 23, 1999).

American Association of Occupational Health Nurses. 1999b. *Compensation and Benefits Study, 2nd ed*. Atlanta: American Association of Occupational Health Nurses, Inc.

American Association of Occupational Health Nurses. 1999c. Proposed competencies (and performance criteria) in occupational and environmental health nursing. Unpublished manuscript. April. American Association of Occupational Health Nurses, Atlanta.

American Board of Industrial Hygiene. 1998. Definitions and functions of industrial hygiene. *Bulletin of the American Board of Industrial Hygiene* [www document]. URL www.den.davis.ca.us/go/abih/bulletin.htm#article3. (accessed on July 15, 1999).

American Board of Preventive Medicine. 1999. Booklet of information. [www document]. URL http://abprevmed.org/infobook.htm. (accessed May 3, 1999).

American College of Occupational and Environmental Medicine. 1999a. Membership Information. [www document]. URL http://www.acoem.org/member/member.htm (accessed May 3, 1999).

American College of Occupational and Environmental Medicine. 1999b. *Annual Report, 1998–1999*. Arlington Heights, IL: American College of Occupational and Environmental Medicine.

American College of Occupational and Environmental Medicine. 1999c. Demographic Profile Members Questionnaire, October, 1999. Arlington Heights, IL: American College of Occupational and Environmental Medicine.

American College of Preventive Medicine. 1999. Preventive medicine residency programs. [www document]. URL http://acpm.org/res_prog.htm (accessed May 28, 1998).

American Conference of Governmental Industrial Hygienists. 1999. Profile of membership. [www document]. URL http://www.acgih.org/members/profile.htm (accessed March 16, 1999).

American Industrial Hygiene Association. 1994. *The American Industrial Hygiene Association: Its History and Personalities*. Fairfax, VA: American Industrial Hygiene Association.

American Management Association. 1999. Workplace Monitoring and Surveillance: Summary of Key Findings. [www document]. URL www.amanet.org/research/specials/compendium.htm#HRI. (accessed on December 12, 1999).

American Society of Safety Engineers and the Board of Certified Safety Professionals. 1991. *Curriculum Standards for Baccalaureate Degrees in Safety*. Des Plaines: American Society of Safety Engineers and the Board of Certified Safety Professionals

American Society of Safety Engineers and the Board of Certified Safety Professionals. 1994a. *Curriculum Standards for Master's Degrees in Safety*. Des Plaines, IL: American Society of Safety Engineers and the Board of Certified Safety Professionals

American Society of Safety Engineers and the Board of Certified Safety Professionals. 1994b. *Curriculum Standards for Safety Engineering, Master's Degrees in Safety, and Safety Engineering Options in Other Engineering Master's Degrees*. Des Plaines, IL: American Society of Safety Engineers and the Board of Certified Safety Professionals

American Society of Safety Engineers and the Board of Certified Safety Professionals. 1995. *Curriculum Standards for Associate Degrees in Safety*. Des Plaines, IL: American Society of Safety Engineers and the Board of Certified Safety Professionals.

American Society of Safety Engineers and Board of Certified Safety Professionals. 1997. *Career Guide to the Safety Profession*. Des Plaines, IL: American Society of Safety Engineers.

American Society of Safety Engineers. 1996. *Scope and Functions of the Professional Safety Position*. Des Plaines, IL: American Society of Safety Engineers.

American Society of Safety Engineers. 1997. *Membership Analysis, Preliminary*. Des Plaines, IL: American Society of Safety Engineers.

American Society of Safety Engineers. 1999. *1998–1999 College and University Survey*. Des Plaines, IL: American Society of Safety Engineers.

Amirault, T. 1997. Characteristics of multiple jobholders, 1995. *Monthly Labor Review* 120(3):9–15.

Angell, M., and J. P. Kassirer. 1996. Quality and the medical marketplace—following elephants. *New England Journal of Medicine* 335:883–885.

Anstadt, G. 1999. Presentation to Committee to Training Needs for Occupational Safety and Health Personnel in the United States, Institute of Medicine, Washington, DC, May 26, 1999.

Association Research Inc. 1997. *Summary of the Results, Definition of the Profession Survey April 1997*. Rockville, MD: Association Research Inc.

Atkinson, G. 1998. Political economy of liberalization and regulation: trade policy for the new era. *Journal of Economic Issues* 32:419–426.

Babbitz, M. 1983. The practice of occupational health nursing in the U.S. *Occupational Health Nursing* 31:23–25.

Barker, K. 1995. Contingent work: research issues and the lens of moral exclusion, pp. 31–60. *In: Changing Employment Relations: Behavioral and Social Perspectives*, L. E. Tetrick and J. Barling, eds. Washington, DC: American Psychological Association.

Bendiner, E. 1995. Alice Hamilton's war on occupational disease. *Hospital Practice* May:80–88.

Bernard, B. P. ed. 1997. *Musculoskeletal Disorders and Workplace Factors*. NIOSH Publication No. 97-141. Cincinnati: National Institute for Occupational Safety and Health.

Blumenthal, D. 1999. Health care reform at the close of the 20th century. *New England Journal of Medicine* 340:1916–1920.

Bodenheimer, T. 1999. The American health care system. The movement for improved quality in health care. *New England Journal of Medicine* 340:488–492.

Bodenheimer, T., and K. Sullivan. 1998. How large employers are shaping the health care marketplace. *New England Journal of Medicine* 338:1003–1007, 1084–1087.

Bronstein, N. E. 1998. *The Chance to Help Whole Populations at Risk: Occupational Physicians Scholarship Fund 1998 Report*. Schiller Park, IL: Occupational Physicians Scholarship Fund.

Bureau of Economic Analysis. 1999. *Survey of Current Business (September)*. Washington, DC: Bureau of Economic Analysis.

Bureau of Health Professions. 1997. *The Registered Nurse Population: March 1996 Findings from the National Sample Survey of Registered Nurses*. Rockville, MD: Health Resources and Services Administration.

Bureau of Labor Statistics. 1995. Contingent worker and alternative employment. [www document] URL www.bls.census.gov/cps/contwkr/contwkr.htm. (accessed December 12, 1999).

Bureau of Labor Statistics. 1996. *Issues in Labor Statistics, Older Workers Injuries Entail Lengthy Absences from Work*. Bureau of Labor Statistics Summary 96-6, April. Washington, DC: U.S. Department of Labor.

Bureau of Labor Statistics. 1997. Employment outlook: 1996–2006. *Monthly Labor Review* 120(11):1–83.

Bureau of Labor Statistics. 1998a. *Occupational Injuries and Illnesses in U.S. Industry, 1997*. Washington, DC: U.S. Department of Labor.

Bureau of Labor Statistics. 1998b. *Issues in Labor Statistics, Women Experience Fewer Job-Related Injuries and Deaths than Men*. Bureau of Labor Statistics Summary 98-8, July. Washington, DC: U.S. Department of Labor.

Bureau of Labor Statistics. 1998c. *Employment and Wages, Annual Averages*. BLS Bulletin 2511 (December). Washington, DC: U.S. Department of Labor.

Bureau of Labor Statistics. 1998d. Work at Home in 1997, Current Population Survey. Press release 98–93 (March). Washington, DC: U.S. Department of Labor.

Bureau of Labor Statistics. 1998e. *National Longitudinal Survey*. Washington, DC: U.S. Department of Labor.

Bureau of Labor Statistics. 1998f. Workers on flexible and shift schedules in 1997. Press Release 98-119 (March). Washington, D.C: U.S. Department of Labor.

Bureau of Labor Statistics. 1999a. Survey of Occupational Injuries and Illnesses, *Monthly Labor Review* 122(8):94.

Bureau of Labor Statistics. 1999b. *Fatal Workplace Injuries in 1997: A Collection of Data and Analysis* Report 934 (Table A-4, pages 64–73). Washington, DC: U.S. Department of Labor.

Burke, R. J., and D. Nelson. 1998. Mergers and acquisitions, downsizing, and privatization: a North American perspective, pp. 21–54. *In: The New Organizational Reality: Downsizing, Restructuring, and Revitalization*. M. K. Gowing, J. D. Kraft, and J. C. Quick, eds. Washington, DC: American Psychological Association.

Burnfield, J. L., and G. J. Medsker. 1999. Income and employment of SIOP members in 1997. *TIP: The Industrial-Organizational Psychologist* 36(4) [www document] URL http://www.siop.org/tip/backissues/Tipapr99/Tipapr99.htm (accessed July 30, 1999).

Burton, J. 1996. Workers compensation twenty-four hour coverage and managed care. *John Burton's Workers' Compensation Monitor* 9(1):11–22.

Business Software Alliance. 1999. Forecasting a robust future: An economic study of the U.S. software industry. [www document] URL http://www.bsa.org/statistics/ (accessed September 1, 1999).

Cascio, W. F. 1993. Downsizing: What do we know? What have we learned? *Academy of Management Executive* 7:95–104.

Cassinelli, M. E., and P. F. O'Connor (Eds.) 1994. *NIOSH Manual of Analytic Methods*. DHHS Publication 94-113. Washington, DC: U.S. Government Printing Office.

Castorina, J. S., and L. Rosenstock. 1990. Physician shortage in occupational and environmental medicine. *Annals of Internal Medicine* 113:983–986.

Caudron, S. 1992. Working at home pays off. *Personnel Journal*, November, pp. 40–49.

Centers for Disease Control and Prevention. 1999. Achievements in public health, 1900–1999: improvements in workplace safety—United States, 1900–1999. *Morbidity and Mortality Weekly Reports* 48(22):461–469.

Chassin, M. R., R. W. Galvin, and the National Roundtable on Health Care Quality. 1998. The urgent need to improve health care quality. *Journal of the American Medical Association* 280:1000–1005.

Chiswick, B. R., and P. W. Miller. 1998. English language proficiency among immigrants in the United States. *Research in Labor Economics* 17:48–51.

Christian, J. In Press. Reducing lost work days: a mostly non-medical matter. *Journal of Workers Compensation*.

Cohen, A., and M. J. Colligan. 1998. *Assessing Occupational Safety and Health Training: A Literature Review*. Cincinnati: National Institute for Occupational Safety and Health.

Cone, J. E., A. Dapone, D. Makofsky, R. Reiter, C. Becker, R. J. Harrison, and J. Balmes. 1991. Fatal injuries at work in California. *Journal of Occupational Medicine* 33:813–817.

Consumer Reports. 2000. Workers comp: falling down on the job. *Consumer Reports*, February, pp. 28–33.

Cooper, R., P. Laud, and C. Dietrich. 1998a. Current and projected workforce of nonphysician clinicians. *Journal of the American Medical Association* 280:788–794.

Cooper, R., T. Henderson, and C. Dietrich. 1998b. Roles of nonphysician clinicians as autonomous providers of patient care. *Journal of the American Medical Association* 280:795–802.

Cooper, S. F. 1995. The expanding use of the contingent workforce in the American economy: new opportunities and dangers for employers. *Employee Relations Law Journal* 20:525–538.

Coovert, M. D. 1995. Technological changes in office jobs: what we know and what we can expect, pp. 175–208. In: *The Changing Nature of Work*, A. Howard, ed. San Francisco: Jossey-Bass.

Corn, J. K. 1992. *Response to Occupational Health Hazards: A Historical Perspective*. New York: Nostrand Reinhold.

Corn, M., and J. Corn, eds. 1988. *Training and Education in Occupational Hygiene: An International Perspective*. Cincinnati, OH: American Council of Government Industrial Hygienists.

Cox, A. 1999. Presentation to Committee to Assess Training Needs for Occupational Safety and Health Personnel, in the United States, Institute of Medicine. Washington, DC, May 26, 1999.

Cox, J. L., and W. Johnston. 1985. A Study of the Impact of Occupational Safety and Health Training and Education Programs on the Supply and Demand for Occupational Safety and Health Professionals. Cincinnati, OH: National Institute for Occupational Safety and Health.

D'Aunno, T. 1996. Business as usual? Changes in health care's workforce and organization of work. *Hospital and Health Services Administration* 41:3–18.

Dembe, A., K. M. Rest, and L. Rudolph. 1998. *The role of prevention in workers' compensation managed care agreements. Occupational Medicine: State of the Art Reviews* 13(4):663–677.

Denison, D., and A. Mishra. 1995. Toward a theory of organizational culture and effectiveness. *Organization Science* 6:204–223.

de Vries, M. F. R. K., and K. Balasz. 1997. The downside of downsizing. *Human Relations* 50:11–50.

Ducatman, A. M. 1986. Workers' compensation cost-shifting. a unique concern of providers and purchasers of prepaid health care. *Journal of Occupational Medicine* 28(11):1174–1176.

Eccleston, S. M. 1995. *Managed care and Medical Cost Containment in Workers' Compensation: A National Inventory 1995–1996.* Cambridge, MA: Worker's Compensation Institute.

Eisenberg, D., R. Davis, S. Ettner, S. Appel, S. Wilkey, M. Van Rompay, and R. Kessler. 1998. Trends in alternative medicine use in the United States, 1990–1997. *Journal of the American Medical Association* 280:1569–1575.

Ellrodt, G., D. Cook, J. Lee, D. Hunt, and S. Weingarten. 1997. Evidence-based disease management. *Journal of the American Medical Association* 278:1687–1692.

Employee Assistance Professionals Association. 1999. *EAPA Needs Assessment.* Arlington, VA: Employee Assistance Professional Association.

Ewell, M., and K.O. Ha. 1999. Piecework practices curtailed. *San Jose Mercury News,* October 17, 1999.

Fair, G. E. 1998. Don't layoff, salvage and retain. *Employee Assistance Quarterly* 13:61–75.

Fierman, J. 1994. The contingency workforce. *Fortune,* January 24, pp. 30–34.

Filipczak, R. et al. 1995. Contingent worker numbers will grow. *Training* 32:11–12.

Fielding, J. 1994. Where is the health in health system reform? *Journal of the American Medical Association* 272:1292–1296.

Frank, A.L. 1999. Ethical aspects of genetic testing, *Mutation Research* 428:285–290.

Fronstin, P. 1998. *Features of Employer Based Health Plans.* Washington, DC: Employee Benefit Research Institute.

Frumpkin, H. 1998. Free-Trade Agreements. *In: Encyclopaedia of Occupational Health and Safety,* 4th ed., Stellman, J. M., ed. Geneva: International Labour Office.

Fuchs, V. R. 1997. Managed care and merger mania. *Journal of the American Medical Association* 277:920–921.

Galinsky, E., and J. T. Bond. 1998. *The 1998 Business Worklife Study.* New York: Families and Work Institute.

The Gary Siegal Organization, Inc. 1996. ACOEM Membership Survey: Scope of Practice. Lincolnwood, IL: The Gary Siegal Organization.

Ginzberg, E. 1995. A cautionary note on market reforms in health care. *Journal of the American Medical Association* 274:1633–1634.

Godfrey, H. 1978. One hundred years of industrial nursing. *Nursing Times* 74(48):1966–1969.

Goetzel, R., D. Anderson, W. Whitmer, R. J. Ozminkowski, R. L. Dunn, J. Wasserman, and Health Enhancement Research Organization Research Committee. 1998. The relationship between modifiable health risks and health care expenditures: An analysis of the multi-employer HERO health risk and cost database. *Journal of Occupational and Environmental Medicine* 40(10):500–510.

Goldman R.H., S. Rosenwasser, and E. Armstrong. 1999. Incorporating an environmental/occupational medicine theme into the medical school curriculum. *Journal of Occupational and Environmental Medicine* 41:47–52.

Greaves, W. W. 1999. Presentation to Committee to Assess Training Needs for Occupational Safety and Health Personnel, in the United States, Institute of Medicine, Washington, DC, May 26.

Green-McKenzie, J., J. Parkerson, and E. Bernacki. 1998. Comparison of workers' compensation costs for two cohorts of injured workers before and after the introduction of managed care. *Journal of Occupational and Environmental Medicine* 40(6):568–572.

Hackett, E. J., P. H. Mirvis, and A. L. Sales. 1991. Women's and men's expectations about the effects of new technology at work. *Group and Organization Studies* 16:60–85.

Hale, A. R. 1984. Is safety training worthwhile? *Journal of Occupational Accidents* 6:17–23.

Hamilton, A. 1990. The Illinois survey: exploring the dangerous trades. *Harvard Medical,* Summer, pp. 37–41.

Hashimoto, D. 1996. Defining the role of managed care in workers' compensation. *Occupational Medicine: State of the Art Reviews* 11(1):101–112.

Heaney, C. A., and R. Z. Goetzel. 1997. A review of health-related outcomes of multi-component worksite health promotion programs. *American Journal of Health Promotion* 11(4):290–307.

Hofmann, D. A., and A. Stetzer. 1996. A cross-level investigation of factors influencing unsafe behaviors and accidents. *Personnel Psychology* 49:307–339.

Howard, A. 1995. Rethinking the psychology of work, pp. 513–555. *In: The Changing Nature of Work*, A. Howard, ed. San Francisco: Jossey-Bass.

Human Resources Institute. 1996. *The Contingent Workforce*. St. Petersburg, FL: Eckerd College

Iglehart, J. 1999a. The American health care system. Expenditures. *New England Journal of Medicine* 340:70–76.

Iglehart, J. 1999b. The American health care system. Medicare. *New England Journal of Medicine* 340:327–332.

Iglehart, J. 1999c. The American health care system. Medicaid. *New England Journal of Medicine* 340:403–408.

Institute of Medicine. 1988. *Role of the Primary Care Physician in Occupational and Environmental Medicine*. Washington, DC: National Academy Press.

Institute of Medicine. 1991. *Addressing the Physician Shortage in Occupational and Environmental Medicine*. Washington, DC: National Academy Press.

Institute of Medicine. 1996. *Primary Care: America's Health in a New Era*. Washington, DC: National Academy Press.

Institute of Medicine. 1998. *Protecting Youth at Work*. Washington, DC: National Academy Press.

Institute of Medicine. 1999. *Reducing the Burden of Injury*. Washington, DC: National Academy Press.

Integrated Data Corporation. 1998. *Tracker,* Fourth Quarter.

Interstudy. 1998. *The Interstudy HMO Trend Report, 1987–1997*. St. Paul, MN: Interstudy.

Jacoby, S. M. 1985. *Employing Bureaucracy: Managers, Unions, and the Transformation of Work in American Industry, 1990–1945*. New York: Columbia University Press.

Jones, R., J. Ganem, D. Williams, and J. Krakower. 1998. Review of U.S. medical school finances, 1996–1997. *Journal of the American Medical Association* 280:813–818.

Kaptchuk, T., and D. Eisenberg. 1998. The persuasive appeal of alternative medicine. *Annals of Internal Medicine* 129:1061–1065.

Karasek, R., and D. Theorell. 1990. *Health Work: Stress, Productivity and the Reconstruction of Working Life*. New York: Basic Books.

Keyserling, W. M. 1995. Occupational Safety: Prevention of Accidents and Overt Trauma. *In: Occupational Health: Recognizing and Preventing Work-Related Disease*, B. S. Levy and D. H. Wegman, eds. Boston: Little, Brown and Company.

Kochan, T. A., J. C. Wells, and M. Smith. 1992. Consequences of a failed IR system: Contract workers in the petrochemical industry. *Sloan Management Review*, Summer, pp. 79–89.

Kosinski, M. 1998. Effective outcomes management in occupational and environmental health. *AAOHN Journal* 46(10):500–510.

Kozlowski, S. W. J., G. T. Chao, E. M. Smith, and J. Hedlund. 1993. Organizational downsizing: strategies, interventions, and research implications, pp. 263–332. *In: International Review of Industrial and Organizational Psychology*, C. L. Cooper and I. T. Robertson (eds.)

Kuttner, R. 1999a. The American health care system: health insurance coverage. *New England Journal of Medicine* 340:163–168.

Kuttner, R. 1999b. The American health care system. Wall Street and health care. *New England Journal of Medicine* 340:664–668.

Landsbergis, P., J. Cahill, and P. Schnall. 1999. The impact of lean production and related new systems of work organization on worker health. *Journal of Occupational Health Psychology* 4:108–130.

Leigh, J. P., S. B. Markowitz, M. C. Fahs, C. Shin, and P. J. Landrigan. 1997. Occupational injury and illness in the United States: estimates of costs, morbidity, and mortality. *Archives of Internal Medicine* 157:1557–1568.

Leigh, J. P., and T. Miller. 1998. Occupational illnesses within two national data sets. *International Journal of Occupational Environmental Health* 4:99–113.

Leone, F. H., and K. J. O'Hara. 1998. The market for occupational medicine managed care. *Occupational Medicine: State of the Art Reviews* 13:869–879.

Levenstein, C. 1988. Occupational Health in the United States: an overview. *In: Occupational Health, Recognizing and Preventing Work-Related Disease*, 2nd ed., B. S. Levy and D. H. Wegman, eds. New York: Little, Brown and Company.

Levit, K., C. Cowan, B. Braden, J. Stiller, A. Sensenig, and H. Lazenby. 1998. National health expenditures in 1997: more slow growth. *Health Affairs* 17:99–110.

Lohr, K., ed. 1990. *Medicare: A Strategy for Quality Assurance*. Washington, DC: National Academy Press.

Martin, R. E., and S. J. Freeman. 1998. The economic context of the new organizational reality, pp. 21–54. *In: The New Organizational Reality: Downsizing, Restructuring, and Revitalization*, M. K. Gowing, J. D. Kraft, and J. C. Quick, eds. Washington, DC: American Psychological Association.

Meyer, G. S., A. Potter, and N. Gary. 1997. A national survey to define a new core curriculum to prepare physicians for managed care practice. *Academic Medicine* 72(8):669–676.

Mohrman, S. A., and S. G. Cohen. 1995. When people get out of the box: new relationships, new systems, pp. 365–410. *In: The Changing Nature of Work*, A. Howard, ed. San Francisco: Jossey-Bass.

Mont, D., J. F. Burton, Jr., and V. Rens. 1999. *Workers' Compensation: Benefits, Coverage, and Costs, 1996*. Washington, DC: National Academy of Social Insurance.

Moon, S. D., and S. L. Sauter, eds. 1996. *Beyond Biomechanics: Psychosocial Aspects of Musculoskeletal Disorders in Office Work*. London: Taylor and Francis.

National Center for Education Statistics. 1993. *Adult Literacy in America*. Washington, DC: U.S. Department of Education.

National Center for Education Statistics. 1997. *Distance Education in Higher Education Institutions*. Statistical Analysis Report NCES 98-062. Washington, DC: U.S. Department of Education.

National Clearinghouse for Worker Safety and Health Training. 1999. *Safety and health resources*. [www document] URL http://204.177.120.20/index.htm (accessed August 12, 1999).

National Institute of Environmental Health Sciences. 1999. *Hazwoper Training: Utilizing Advanced Training Technologies*. Report of a Technical Workshop on Computer and Internet-based Learning Methods for Safety and Health Training, conducted April 20–21, Estes Park, Colorado. Research Triangle, NC: National Institute of Environmental Health Sciences.

National Institute for Occupational Safety and Health. 1978. *A Nationwide Survey of the Occupational Safety and Health Work Force*. Cincinnati: National Institute for Occupational Safety and Health.

National Institute for Occupational Safety and Health. 1988. *Annual Statistical Report of NIOSH Training Grant Outputs*. Cincinnati: National Institute for Occupational Safety and Health.

National Institute for Occupational Safety and Health. 1989. *Annual Statistical Report of NIOSH Training Grant Outputs*. Cincinnati: National Institute for Occupational Safety and Health.

National Institute for Occupational Safety and Health. 1990. *Annual Statistical Report of NIOSH Training Grant Outputs*. Cincinnati: National Institute for Occupational Safety and Health.

National Institute for Occupational Safety and Health. 1991. *Annual Statistical Report of NIOSH Training Grant Outputs*. Cincinnati: National Institute for Occupational Safety and Health.

National Institute for Occupational Safety and Health. 1992. *Annual Statistical Report of NIOSH Training Grant Outputs*. Cincinnati: National Institute for Occupational Safety and Health.

National Institute for Occupational Safety and Health. 1993. *Annual Statistical Report of NIOSH Training Grant Outputs*. Cincinnati: National Institute for Occupational Safety and Health.

National Institute for Occupational Safety and Health. 1994. *Annual Statistical Report of NIOSH Training Grant Outputs*. Cincinnati: National Institute for Occupational Safety and Health.

National Institute for Occupational Safety and Health. 1995. *Annual Statistical Report of NIOSH Training Grant Outputs*. Cincinnati: National Institute for Occupational Safety and Health.

National Institute for Occupational Safety and Health. 1996. *Annual Statistical Report of NIOSH Training Grant Outputs*. Cincinnati: National Institute for Occupational Safety and Health.

National Institute for Occupational Safety and Health. 1997a. *Annual Statistical Report of NIOSH Training Grant Outputs*. Cincinnati: National Institute for Occupational Safety and Health.

National Institute for Occupational Safety and Health. 1997b. *Pocket Guide to Chemical Hazards*. Washington, DC: U.S. Government Printing Office.

National Institute for Occupational Safety and Health. 1998a. *National Occupational Research Agenda Update*. Cincinnati: National Institute for Occupational Safety and Health.

National Institute for Occupational Safety and Health. 1998b. *Annual Statistical Report of NIOSH Training Grant Outputs*. Cincinnati: National Institute for Occupational Safety and Health.

National Institute for Occupational Safety and Health. 1999. *Identifying High-Risk Small Business Industries*. Publication No. 99-107. Cincinnati: National Institute for Occupational Safety and Health.

National Research Council. 1994. *Information Technology in the Service of Society: A Twenty-First Century Lever*. Washington, DC: National Academy Press.

National Research Council. 1995. *Information Technology for Manufacturing: A Research Agenda*. Washington, DC: National Academy Press.

National Research Council. 1999. *The Changing Nature of Work: Implications for Occupational Analysis*. Washington, DC: National Academy Press.

National Safety Council. 1999. Injury Facts 1999. [www document] URL http:// www.nsc.org/library.htm. (accessed December 14, 1999).

Nikolaj, S., and B. Boon. 1998. Health care management in workers' compensation. *Occupational Medicine: State of the Art Reviews* 13(2):357–359.

North American Free Trade Agreement. 1994. NAFTA. [www document] URL http:// www.treas.gov/press/releases/docs/nafta.htm (accessed October 4, 1999).

Occupational Safety and Health Administration. 2000. OSHA Facts. [www document] URL http://www.osha-slc.gov/OSHA Facts/OSHA Facts.html (accessed March 9, 2000).

Occupational Safety and Health Administration. 1999a. Current Susan Harwood Training Grants. [www document] URL http://www.osha-slc.gov/Training/sharwood/ current.html (accessed August 9, 1999).

Occupational Safety and Health Administration. 1999b. OSHA Outreach Training Program. [www document] URL http://www.osha-slc.gov/Training/Bulletin/index.html (accessed August 9, 1999).

Occupational Safety and Health Administration. 1999c. State Occupational Safety and Health Plans. [www document] URL http://www.osha.gov/oshprogs/stateprogs. html (accessed July 23, 1999).

Occupational Safety and Health State Plan Association. 1999. Grassroots worker protection: 1999 report. [www document] URL http://www.wa.gov/lni/wisha/grwp.htm (accessed July 26, 1999).

Office of Inspector General. 1996. Center for Disease Control and Prevention's Educational Resource Centers. Atlanta: Department of Health and Human Services.

Office of the President. 1999. *Economic Report of the President* (Table B-3). Washington, DC: Office of the President.

Office of Technology Assessment. 1985. *Preventing Injury and Illness in the Workplace*. Report OTA-H-256. Washington, DC: U.S. Government Printing Office.

Office of Technology Assessment. 1991. *Biological Rhythms: Implications for the Worker*. Report OTA-BA-463. Washington, DC: U.S. Government Printing Office.

Office of Training and Education. 1996. *Student Monthly Report for September 1996*. Des Plaines, IL: Occupational Safety and Health Administration Training Institute.

Office of Training and Education. 1997. *Student Monthly Report for September 1997*. Des Plaines, IL: Occupational Safety and Health Administration Training Institute.

Office of Training and Education. 1998. *Student Monthly Report for September 1998*. Des Plaines, IL: Occupational Safety and Health Administration Training Institute.

Oliver, N., and B. Wilkinson. 1992. *The Japanization of British Industry: New Developments in the 1990s*. Oxford: Blackwell.

Palmer, I., B. Kabanoff, and R. Dunford. 1997. Managerial accounts of downsizing. *Journal of Organizational Behavior* 18:623–639.

Parker, D. L., W. R. Carl, L. R. French, and F. B. Martin. 1994. Characteristics of adolescent work injuries reported to the Minnesota Department of Labor and Industry. *American Journal of Public Health* 84:606–611.

Pearson, R. J., W. M. Kane, and H. K. Keimowitz. 1988. The preventive medicine physician: A national study. *American Journal of Preventive Medicine* 4:289–297.

Pearlstein, S. 1994. Large U.S. companies continue downsizing. *The Washington Post*, September 27, p. C1.

Pelletier, K. R. 1996. A review and analysis of the health case cost-effective outcome studies of comprehensive health promotion and disease prevention programs at the worksite: 1993-1995 update. *American Journal of Health Promotion* 10:380–388.

Peterson, J. E. 1999. Past President's Column: Functions of the American Academy of Industrial Hygiene. *AAIH Newsletter*, March.

Pope, A. M., and D. P. Rall (Eds.) 1995. *Environmental Medicine: Integrating a Missing Element into Medical Education*. Washington, D.C.: National Academy Press.

Proehl, R. A. 1996. Cross-functional teams: A panacea or just another headache? *Supervision* 57(7):6.

Quick, J. D. 1999. Occupational health psychology: historical roots and future directions. *Health Psychology* 18:82–88.

Quick, J. C., L. R. Murphy, and J. J. Hurrell, Jr. 1992. *Stress and Well-Being at Work*. Washington, DC: American Psychological Association.

Reuter, J., and D. Gaskin. 1997. Academic health centers in competitive markets. *Health Affairs* 16:242–252.

Rivo, M. L., H. L. Mays, J. Katzoff, and D. A. Kindig. 1995. Managed health care: implications for the physician workforce and medical education. *Journal of the American Medical Association* 274(9):712–715.

Rogers, B. 1988. Perspectives in occupational health nursing. *AAOHN Journal* 36:151–155.

Rogers, B. 1994. *Occupational Health Nursing: Concepts and Practice*. Philadelphia: W.B. Saunders & Co.

Rogers, B. 1998. Occupational health nursing expertise. *AAOHN Journal* 46:447–483.

Rom, W. N. 1992 The Discipline of Environmental and Occupational Medicine. In: *Environmental and Occupational Medicine*, 2nd ed., W. N. Rom, ed. Philadelphia: Lippencott-Raven.

Rousseau, D. M., and K. A. Wade-Benzoni. 1995. Changing individual-organization attachments: a two-way street. In: *The Changing Nature of Work*, A. Howard, ed. San Francisco: Jossey-Bass.

Rundall, T. 1994. The integration of public health and medicine. *Frontiers of Health Services Management* 10:3–24.

Sauter, S. L., J. J. Hurrell, H. R. Fox, L. E. Tetrick, and J. Barling. 1999. Occupational health psychology: an emerging discipline. *Industrial Health* 37:199–211.

Schiff, G. 1996. 12 Fundamental problems. Why for-profit managed care fails you and your patients. *ACP Observer*, November, pp. 13–15.

Schnall, P. L., J. E. Schwartz, P. A. Landsbergis, K. Warren, and T. G. Pickering. 1998. A longitudinal study of job strain and ambulatory blood pressure: results from a three-year follow-up. *Psychosomatic Medicine* 60:697–706.

Schneider, B. 1987. The people make the place. *Personnel Psychology* 49:307–339.

Schoen, C., C. Hoffman, D. Rowland, K. Davis, and D. Altman. 1998. *Working Families at Risk: Coverage, Access, Cost, and Worries*. New York: Commonwealth Fund.

Schroeder, S. A. 1996. The medically uninsured—will they always be with us? *New England Journal of Medicine* 334:1130–1133.

Shaiken, H. 1985. *Work Transformed*. New York: Holt, Rinehart, and Winston.

Small Business Administration Office of Advocacy. 1998. *Small Business Growth by Major Industry, 1988–95*. Washington, DC: U.S. Small Business Administration.

Smith, S., M. Freeland, S. Heffler, and D. McKusick. 1998. The next ten years of health spending: what does the future hold? The Health Expenditures Projection Team. *Health Affairs* 17:128–140.

Smyth, H. F., Jr. 1966. Certification in occupational health: philosophy, implication, and mechanism. II. The American Board of Industrial Hygiene. *American Journal of Public Health* 56:1120–1127.

Solomon, B. 1993. Using managed care to control workers' compensation costs. *Compensation and Benefits Review* 25:59–65.

Sparks, P. J., and A. Feldstein. 1997. The success of the Washington department of labor and industries managed care pilot project: the occupational medicine-based delivery model. *Journal of Occupational and Environmental Medicine* 39(11):1068–1073.

Stephens, T. 1999. Distance education. Unpublished paper commissioned by the Committee to Assess Training Needs for Occupational Safety and Health Personnel in the United States, Institute of Medicine, Washington, DC.

Stinson, J., Jr. 1997. New data on multiple jobholding available from CPS. *Monthly Labor Review* 120(3):3–8.

Stout, N. A., and C. A. Bell. 1991. Effectiveness of source documents at identifying fatal occupational injuries: a synthesis of studies. *American Journal of Public Health* 81:725–728.

Tan, K. K., N. G. Fishwick, W. A. Dickson, and P. J. Sykes. 1991. Does training reduce the incidence of industrial hand injuries? *Journal of Hand Surgery* 16B:323–326.

Training Magazine. 1997. Industry report. *Training Magazine* 34:33–75.

Upfal, M. and W. Shaw, and The American College of Occupational and Environmental Medicine Panel to Define the Competencies of Occupational and Environmental Medicine. 1998. Occupational and environmental medicine competencies-V1.0. *Journal of Occupational and Environmental Medicine* 40:427–440.

U.S. Census Bureau. 1998. *Health Insurance Coverage: 1998.* Washington, DC: U.S. Government Printing Office.

U.S. Department of Commerce. 1999. Computer Equipment Industry Trends and Forecasts [www document] URL http://infoserve2.ita.doc.gov/ocbe/USIndust.nsf/439d5384 b275b29a8525653400455415/62f44f0f17a41b448525651600692b35?OpenDocument

U.S. Department of Justice. 1999. ADA homepage. [www document] URL http://www.usdoj.gov/crt/ada/adahom1.htm Accessed November 20, 1999.

U.S. Department of Justice. 1998. Workplace violence, 1992–96. [www document] URL http://www.ojp.usdoj.gov/bjs/pub/ascii/wv96.txt Accessed February 10, 2000.

U.S. Department of Labor. 1999a. Technology and Globalization. In: *Futurework: Trends and Challenges for Work in the 21st Century.* Washington, DC: U.S. Department of Labor.

U.S. Department of Labor. 1999b. *Report on the American Workforce.* Washington, DC: U.S. Department of Labor.

Wall, T. D., and P. R. Jackson. 1995. New manufacturing initiatives and shopfloor job design. In: *The Changing Nature of Work,* A. Howard, ed. San Francisco: Jossey-Bass.

Weinper, M. 1999. Workers' comp gets managed. *RehabManagement* December/January:34–36.

Wellins, R. S., W. C. Byham, and J. M. Wilson. 1991. *Empowered Teams: Creating Self-Directed Work Groups That Improve Quality, Productivity, and Participation.* San Francisco: Jossey-Bass Publishers.

Whitehead, L. W., and M. S. West. 1997. CIH+IHIT utilization by industry or industry group, and preliminary projections of future need for such IH professionals. Paper presented at American Industrial Hygiene Conference and Exposition, Dallas, May 19, 1997.

Zikiye, A. A., and R. A. Zikiye. 1992. Satisfaction gaps: New realities in managing automation. *Management Decision* 30(2):40–45.

Zohar, D. 1980. Safety climate in industrial organizations: theoretical and applied implications. *Journal of Applied Psychology* 65:96–102.

Appendixes

A

Committee and Staff Biographies

LINDA HAWES CLEVER, M.D., M.A.C.P., is founding Chair of the Department of Occupational Health at California Pacific Medical Center. She received undergraduate and medical degrees from Stanford University, undertook further training at Stanford and the University of California, San Francisco, and is board certified in internal medicine and occupational medicine. She is a member of the Western Association of Physicians and is a clinical professor of medicine at the University of California, San Francisco. Her areas of special interest include personal, professional, and organizational renewal; current issues in health care, including managed care and ethics; the interactions of life, work, and health; the occupational health of women and health care workers; and leadership. She has been active in a variety of health, science, education, and public service endeavors. Editor of *Western Journal of Medicine* from 1990 to 1998, Dr. Clever currently serves on the Board of Scientific Counselors of the National Institute for Occupational Safety and Health. She also chairs the Policy Advisory Council of the School of Public Health at the University of California, Berkeley, and the National Advisory Panel of the Institute for Women and Gender at Stanford. Dr. Clever is a member of the Institute of Medicine.

RUTH HANFT, Ph.D., is an Independent Consultant who gained a Ph.D. from the George Washington University, specializing in health services administration and public finance. Dr. Hanft is a former U.S. Department of Health and Human Services Deputy Assistant Secretary for Health Research, Statistics, and Technology, and, more recently professor at the

George Washington University Department of Health Services Management and Policy. Dr. Hanft has led a team of consultants that assessed the School of Public Health at the University of Minnesota, assisted the University of Kansas Masters of Public Health program in its development and preparation for accreditation, and served as a consultant to the Governor of Kansas's Public Health Improvement Commission. Dr. Hanft was also a senior research associate at the Association of Academic Health Centers and served as an adjunct professor in the Department of Community Medicine at Dartmouth University's Medical School. She has won the Walter Patenge Medal of Public Service, awarded by Michigan State University, and was awarded a fellowship by the Hastings Center. She is an associate editor for the *International Journal of Technology Assessment in Health Care*. Dr. Hanft served on the Institute of Medicine Committee on Strategies for Supporting Graduate Medical Training in Primary Care, the Committee on Assessing Research Capabilities in Obstetrics and Gynecology, and the Committee for the Study of Resources for Clinical Investigation. She has been a member of the Institute of Medicine for 21 years.

RONALD KUTSCHER was with the Bureau of Labor Statistics for 39 years and served as Associate Commissioner from 1979 to 1996 where he directed the Office of Economic Growth and Employment Projections. That office was responsible for developing medium-term economic and employment projections of the U.S. economy and for preparing the *Occupational Outlook Handbook*. Mr. Kutscher has extensive international experience, especially in Europe and Asia, where he assisted governments in developing methods for preparing economic projections and assisted them with employment issues, particularly in countries making conversions to market economies. He has published over 50 papers dealing with economic and employment issues. Mr. Kutscher is a currently a fellow of the American Statistical Association and received the Presidential Rank Award in the Senior Executive Service in 1990. Mr. Kutscher received a B.A. in economics from Doane College in Nebraska and has completed graduate studies in economics, statistics, and econometrics at the University of Illinois and at universities in Washington, D.C. He previously served on the Institute of Medicine's Committee on the Adequacy of Nurse Staffing.

JAMES A. MERCHANT, M.D., is Dean of the College of Public Health at the University of Iowa. Dr. Merchant joined the University of Iowa's faculty as a professor, with an academic appointment in preventive medicine in 1981 and an appointment in internal medicine in 1982. He received his M.D. degree from the University of Iowa in 1966 and completed an internship and an internal medicine residency at Cleveland Metropolitan

General Hospital, followed by a fellowship in pulmonary and environmental medicine at Duke University. In 1973 he received a doctorate in public health from the University of North Carolina at Chapel Hill and a Trudeau Fellowship from the American Thoracic Society. His research interests focus on the epidemiology of pulmonary disease, environmental and occupational health, rural health care delivery, agricultural diseases and injuries, international health, and public and rural health policy. He has received numerous grants from National Institute for Occupational Safety and Health, the Centers for Disease Control and Prevention, the National Institutes of Health, and private foundations and corporations. Dr. Merchant currently chairs the Board of Scientific Counselors of the National Institute for Occupational Safety and Health and serves on the National Advisory Committee for Occupational Safety and Health for the U.S. Departments of Labor and Health and Human Services. Among his awards and honors are a commendation medal from the U.S. Public Health Service and a Health Policy Fellowship with the U.S. Senate.

JAMES A. OPPOLD, Ph.D., C.S.P., is an environmental and occupational safety and health consultant, an adjunct professor at East Carolina University, and Chair of the Related Accreditation Commission of the Accreditation Board for Engineering and Technology. He has been a member of the Educational Standards Committee of the American Society of Safety Engineers, a professional/educational standards committee, for 14 years. He holds a doctorate in environmental engineering from the University of Florida as well as a master's degree in radiation biophysics from the University of Kansas. Dr. Oppold previously directed the Occupational Health and Safety Program for the state of North Carolina, served as Director of the Division of Safety Research for National Institute for Occupational Safety and Health at Morgantown, West Virginia, and supervised the radiological health, industrial hygiene, and the safety programs for the Tennessee Valley Authority. Among his consulting clients are the World Health Organization, the International Labor Organization, the World Bank, and the Organization of American States.

M.E. BONNIE ROGERS, Dr.P.H., R.N., F.A.A.N., is an Associate Professor of Nursing and Public Health and Director of the Occupational Health Nursing Program at the University of North Carolina, School of Public Health, Chapel Hill. Dr. Rogers received her nursing degrees from the Washington Hospital Center School of Nursing in Washington, D.C.; and from George Mason University, Fairfax, Virginia. Her master of public health degree and doctor of public health are from the Johns Hopkins University School of Public Health, Baltimore, Maryland. Dr. Rogers is a fellow in the American Academy of Nursing, and the American Association of Occupational Health Nurses. She has several funded research grants on clinical issues in occupational health, research priorities, hazards to health care workers, and ethics. She was a visiting scholar at the Hastings Center in Garrison, New York where she studied ethics and was granted a NIOSH career award to study ethical issues in occupational health. Dr. Rogers has also practiced for many years as a public health nurse, occupational health nurse, and occupational health nurse practitioner. She has published more than 90 articles and book chapters and two books, *Occupational Health Nursing Concepts and Practice,* the only text in the field, and *Occupational Health Nursing Guidelines for Primary Clinical Conditions.* Dr. Rogers serves on numerous editorial boards, is the chairperson of the National Occupational Research Agenda Liaison Committee, and was recently appointed to the National Advisory Committee on Occupational Safety and Health.

SCOTT SCHNEIDER, M.S.I.H., C.I.H., is currently Director of Occupational Health and Safety of the Laborers' Health and Safety Fund of North America, a joint labor management fund affiliated with the Laborers International Union of North America. He was awarded an M.S. in zoology from the University of Michigan in 1973 and an M.S. in industrial hygiene from the University of Pittsburgh in 1980. Mr. Schneider is a Certified Industrial Hygienist, and was Director of the Ergonomics Program at the Center to Protect Workers' Rights, the research arm of the Building Trades department of the American Federation of Labor-Congress of Industrial Organizations. Mr. Schneider is the author of over twenty scholarly works, including chapters in major occupational health and safety reference textbooks, handbooks, and encyclopedias. Mr. Schneider has conducted research for the National Institute for Occupational Safety and Health in areas such as the ergonomics problems of construction workers, construction health hazards, health hazards in an iron foundry, and hearing loss in carpenters and jointers. Among the professional organizations that he belongs to are the American Industrial Hygiene Association, the American Conference of Governmental Industrial Hygienists, and the Human Factors and Ergonomics Society.

MARTIN SEPULVEDA, M.D., F.A.C.P., F.A.C.O.E.M., is the Vice President of Global Occupational Health Services for the International Business Machines Corporation. Dr. Sepulveda is a graduate of Yale University and the Harvard University Medical School. He completed residencies in internal medicine at the University of California-San Francisco Hospitals and Clinics, and in occupational medicine at the National Institute for Occupational Safety and Health. He also completed an advanced fellowship in internal medicine at the University of Iowa Hospitals and Clinics and served as a clinical research investigator at the Division of Respiratory Disease Studies, National Institute for Occupational Safety and Health, and the Centers for Disease Control and Prevention. He is board certified and holds the rank of Fellow in both the American College of Physicians (internal medicine) and the American College of Occupational and Environmental Medicine (occupational medicine). Dr. Sepulveda has published numerous peer-reviewed articles and holds extramural appointments on scientific advisory boards, professional associations, and scientific journal review panels.

ROBERT C. SPEAR, Ph.D., is a Professor of Environmental Health Sciences at the University of California, Berkeley. He received an M.S. in mechanical engineering in 1963 from the University of California, Berkeley and a Ph.D. in control engineering from Cambridge University in Cambridge, England, 5 years later. Dr. Spear's research interests relate principally to the assessment and control of exposures to hazardous agents in both the occupational and community environments. He has an extensive publication record in this field which spans farmworkers' exposures to pesticides to strategies for the characterization and control of the exposure of rural populations to parasites in the developing world. He was appointed founding Director of the University of California's Center for Occupational and Environmental Health in 1979 and continues to serve in that capacity. He has served on numerous committees advisory to governmental agencies including the National Advisory Committee on Occupational Safety and Health and, currently, the National Institute of Occupational Safety and Health Board of Scientific Counselors.

LOIS E. TETRICK, Ph.D., is Professor of Industrial and Organizational Psychology at the University of Houston. Dr. Tetrick is also an Associate Editor of the *Journal of Applied Psychology* and is on the editorial boards of *Journal of Organizational Behavior, Journal of Occupational Health Psychology,* and *Advanced Topics in Organizational Behavior.* She is a fellow of the American Psychological Association, the American Psychological Society, and the Society for Industrial and Organizational Psychology. Before joining the faculty at the University of Houston in 1995, she was on the faculty at Wayne State University for 12 years, where she was the primary mentor of the post-doctoral fellow in occupational health psychology that was funded by the American Psychological Association and National Institute for Occupational Safety and Health. She has conducted research on individuals' perceptions of the employment relationship, employees' commitment to their employers and to their unions, factors associated with occupational stress, and perceptions of the safety climate at work. Professor Tetrick received her doctorate in industrial/organizational psychology from the Georgia Institute of Technology in 1983.

NEAL A. VANSELOW, M.D., is Chancellor-emeritus and Professor-emeritus of Medicine at Tulane University Medical Center. He also holds an appointment as Adjunct Professor of Health Systems Management in the Tulane School of Public Health and Tropical Medicine. He served as Chancellor of Tulane University Medical Center from 1989 to 1994 and as a Scholar-in-Residence at the Institute of Medicine during the 1994 to 1995 academic year. He has served as Chairman of the Department of Post-graduate Medicine and Health Professions Education at the University of Michigan, Dean of the University of Arizona College of Medicine, Chancellor of the University of Nebraska Medical Center, and Vice President for Health Sciences at the University of Minnesota. He is an allergist who received his training in internal medicine and allergy/immunology at the University of Michigan. Dr. Vanselow has served as chairperson of the Council on Graduate Medical Education (U.S. Department of Health and Human Services) and chairperson of the Board of Directors, Association of Academic Health Centers. He has been a member of the Institute of Medicine since 1989 and chaired the Institute of Medicine Committee on the Future of Primary Care. He also served as cochair of the Institute of Medicine Committee on the U.S. Physician Supply. He currently serves as a member of the Pew Health Professions Commission and is on the Board of Trustees of Meharry Medical College. His areas of particular interest include the health care workforce and graduate medical education.

M. DONALD WHORTON, M.D., M.P.H., is an occupational medicine physician certified by the American Board of Internal Medicine, the American Board of Preventive Medicine, and the American College of Epidemiology. Dr. Whorton started his own medical consulting firm, Whorton and Associates, in July 1994. He began his consulting career in 1978 as a Principal and Senior Occupational Physician and Epidemiologist at Environmental Health Associates and remained as the Chief Medical Scientist and a Vice President of ENSR Consulting and Engineering after its purchase in 1988. Dr. Whorton is internationally known for his work on the effects of the nematocide DBCP on the testes of exposed workers and other studies on reproductive effects from workplace exposures. He has authored or co-authored numerous scientific and policy-related articles in professional journals. Dr. Whorton was formerly on the staff of the University of California in Berkeley. A member of many professional organizations, he is a past officer of the American Public Health Association and member of the editorial board of the *American Journal of Public Health* and a past member of the National Institute for Occupational Safety and Health Research Grant Review Study Section. Dr. Whorton is the current chairperson of the University of California State-wide Advisory Committee for Occupational Health Centers. Dr. Whorton is also a member of the Institute of Medicine.

IOM STUDY STAFF

FREDERICK J. MANNING, Ph.D., is a Senior Program officer in the Institute of Medicine's Division of Health Sciences Policy Division and Study Director. In 6 years at Institute of Medicine, he has served as Study Director for projects addressing a variety of topics including medical isotopes, potential hepatitis drugs, blood safety and availability, rheumatic disease, resource sharing in biomedical research, and chemical and biological terrorism. Before joining IOM, Dr. Manning spent 25 years in the U.S. Army Medical Research and Development Command, serving in positions that included Director of Neuropsychiatry at the Walter Reed Army Institute of Research and Chief Research Psychologist for the Army Medical Department. Dr. Manning earned his Ph.D. in psychology from Harvard University in 1970, following undergraduate education at the College of the Holy Cross.

ALDEN B. CHANG II is a project assistant in the Division of Health Sciences Policy. He has been with the Institute of Medicine since February 1999 and has also worked on the *Organ Procurement and Transplantation: Assessing Current Policies and the Potential Impact of the DHHS Final Rule* study and the Forum on Emerging Infections. Alden earned his bachelor of arts degree in international relations from The George Washington University, Washington, D.C.

B
Statement on Committee Composition by Committee Member James A. Oppold

As a member of the Committee to Assess Training Needs for Occupational Safety and Health Personnel in the United States I wish to record my concern about the composition of the 12-member committee and how that influenced the outcome of the report. This statement should be taken as a point of contention and not as a minority objection as I generally support the major conclusions and recommendations. Nevertheless, the summary minutes of the first meeting in March 1999 will show that I voiced concern about insufficient representation of occupational nursing and professional safety on the committee. After some discussion, resolution of this matter was delayed until the end of the first 2-day meeting. After 2 information-packed days I reluctantly agreed that it would be inappropriate to add anyone else to the committee at that time. However, at the November 1999 meeting I again raised the issue that the composition of the committee was far too heavily weighted with occupational physicians. Of the 12 members five are physicians, two are industrial hygienists, one is an occupational health nurse, one is a safety professional, and the other three are non-safety and health professionals: a data-statistician, a psychologist, and a health policy person. For the committee to be more effective I believe it would have been advisable to include safety and health representatives from a business such as insurance, another from engineering, and someone representing the academic programs that teach and do research other than the NIOSH Education and Research Centers.

Another concern is that 50 percent (6) of the present membership of the committee are or have been affiliated with schools of public health or medicine. Their strategies and ideas generally prevailed for a simple majority consensus. I cannot view these decisions as a mandate or even a consensus. For example, the subject of health care, not workers' health care, was an unnecessary topic to be included in the report, but the majority of the committee felt otherwise.

Prevention of injuries, illnesses, and workplace fatalities is a very complex problem and discussion of these problems should have equally included representatives from other disciplines. The discussions would have benefited from ideas and strategies from disciplines such as engineering, business, law, risk analysis, and the sciences (including psychology) and industrial technology. The framers of the Occupational Safety and Health Act of 1970 gave the responsibility for carrying out its directives to two different federal agencies, the U.S. Department of Labor and the U.S. Department of Health, Education and Welfare (now the U.S. Department of Health and Human Services), which indicates that disciplines other than health or more specifically public health should be involved in the prevention of workplace injuries, illnesses, and fatalities.

As for assessing the professional manpower needs the Committee generally agreed that this could not be accomplished within our limited time and research capabilities. Finally, as previously indicated the report could have been better with input from a more balanced committee.

I appreciate the opportunity to serve on the Committee to Assess Training Needs for Occupational Safety and Health Personnel in the United States and trust that my efforts to emphasize injury/fatality prevention brought meaning to the discussions. It is my desire, as it has been for 40 years, to make the place of employment safe and healthy.

C
Significant Events in the History of Occupational Safety and Health

1700   Bernardino Ramazzini, widely considered the "father of industrial medicine," publishes his first book on occupational diseases, *De Morbis Artificum Diatriba (The Diseases of Workmen).*

1812   North America's first accident insurance policy is issued.

1864   The Pennsylvania Mine Safety Act passes into law.

1867   Phillipa Flowerday is hired by the firm of J. & J. Colman in Norwich, Great Britain. Her employment at this mustard company is considered the earliest recorded evidence of a company specifically hiring an industrial nurse.

          The first recorded call by a labor organization for U.S. occupational safety and health law is heard.

          The state of Massachusetts institutes the first government-sponsored factory inspection program.

1888   Betty Moulder of Pennsylvania works with coal miners.

1895   Vermont Marble Company initiates Industrial Nursing Service with Ada Mayo Stewart as the industrial nurse.

1896   The National Fire Protection Association is founded to prevent fires and to write codes and standards.

1897   Great Britain passes a workmen's compensation act for occupational injuries. English legislators would later (1906) extend the aegis of the act to encompass occupational diseases.

1902   The state of Maryland passes the first workers' compensation law.

          The first attempt by a state government to force employers to compensate their employees for on-the-job injuries is overturned when the U.S. Supreme Court declares Maryland's workers' compensation law to be unconstitutional.

1906　First systematic survey of workplace fatalities in the United States is conducted in Allegheny County, Pennsylvania.

1907　Largest coal mining disaster in U.S. history takes place in Monongah, West Virginia.

1908　Alice Hamilton, M.D., the first physician to devote herself to research in industrial medicine, publishes her first article about occupational diseases in the United States.

1911　First U.S. worker's compensation laws are enacted.

A professional, technical organization, the American Society of Mechanical Engineers, responsible for developing safety codes for boilers and elevators, is founded.

National Organization for Public Health Nursing is formed.

1912　National Council for Industrial Safety is established. Originally organized to collect data and promote accident prevention programs, it became the National Safety Council in 1913.

1913　Industrial nurses registry is established in Boston.

The Bureau of Labor Statistics publishes data that show a rate of 61 industrial deaths per 100,000 workers.

1914　The U.S. Public Health Service establishes the Office of Industrial Hygiene and Sanitation. Its primary function is research in occupational health. After several name changes it became the National Institute for Occupational Safety and Health (NIOSH) in 1971.

1916　The U.S. Supreme Court upholds the constitutionality of state workers' compensation laws.

The American Association of Industrial Physicians and Surgeons is formed. It later became the American Occupational Medicine Association, then the American College of Occupational Medicine, and finally, in 1991, the American College of Occupational and Environmental Medicine.

1917　First industrial nursing course is offered at Boston University College of Business Administration.

1918　The American Standards Association is founded. Responsible for the development of many voluntary safety standards, some of which are referenced into laws, today it is known as the American National Standards Institute.

1919　Alice Hamilton, M.D., is appointed assistant professor of industrial medicine at Harvard Medical School, the first woman to be on the faculty of Harvard University.

First book on industrial nursing is written by Florence Wright.

1935   Social Security Act of 1935 is passed. This act provided funds for state industrial programs.

1936   Walsh-Healey Act for worker health and safety standards is enacted, setting safety and health standards for employers receiving federal contracts over $10,000.

1937   Godfrey publishes one of the first statements on the need for public health involvement in accident prevention in the American Journal of Public Health.

The Council on Industrial Health of the American Medical Association is created.

An estimated 2,200 nurses are working in the industry.

1938   American Conference of Governmental Industrial Hygienists is formed.

1939   American Industrial Hygiene Association is formed.

1942   Gordon formalizes concept that epidemiology could be used as a theoretical foundation for accident prevention.

DeHaven describes structural environments as a primary cause of injury in falls from heights.

American Association of Industrial Nurses is founded with Catherine Dempsey as the first President.

1943   Army directives are created for the establishment of industrial medical programs in all Army-owned and operated plants, arsenals, depots, and ports of embarkation.

American Public Health Association Committee on Administrative Practice appoints a subcommittee on accident prevention; the subcommittee reports accident prevention programs in six state and two local health departments.

1946   The American Academy of Occupational Medicine is founded. Its membership comprises full-time physicians in occupational medicine. It merges with the American Occupational Medicine Association in 1988 to form the American College of Occupational and Environmental Medicine.

1948   All states (48 at the time) have workers' compensation laws.

1950   The first doctorates of industrial medicine are conferred upon three graduates of the University of Pittsburgh.

1952   The Coal Mine Safety Act passes into law.

1953   *Human Factors in Air Transportation* is published by McFarland.

Industrial Nursing Journal begun; it later became the *Occupational Health Nursing Journal* and then *AAOHN Journal*.

1955   First annual Stapp conferences on the biomechanics of crashes are held.

American Board on Preventive Medicine recognizes occupational medicine as a subspecialty, with its own certification requirements.

1956 Accident Prevention Program is initiated by the U.S. Public Health Service.

1959 Insurance Institute for Highway Safety is founded.

1960 Specific safety standards are promulgated for the Walsh-Healey Act.

1961 American Public Health Association publishes *Accident Prevention: The Role of Physician and Public Health Workers.*

1964 Passage of the Coal Mine Health and Safety Act greatly expands the powers of federal inspectors. It served as a model for the 1970 Occupational Health and Safety Act.

Journal of Safety Research begins publication.

Haddon, Suchman, and Klein publish *Accident Research: Methods and Approaches.*

Eleven schools of public health develop training programs in injury prevention funded by the U.S. Public Health Service.

The four major U.S. auto manufacturers install front-seat lap belts as standard equipment.

1966 *Accidental Death and Disability: The Neglected Disease of Modern Society* is published by the National Research Council.

The U.S. Department of Transportation and its sections, the National Highway Traffic Safety Administration and the National Transportation Safety Board, are established.

1968 President Lyndon Johnson calls for a federal occupational safety and health law.

1969 Mine Safety and Health Act becomes law.

The Construction Safety Act is passed into law.

Board of Certified Safety Professionals, which certifies practitioners in the safety profession, is established.

Graduate programs in occupational health nursing begin.

1970 Occupational Safety and Health Act is passed into law.

The Occupational Safety and Health Administration and National Institute for Occupational Safety and Health are established.

1972 Black Lung Benefits Act is enacted.

Accreditation Board for Occupational Health Nursing is established.

1974 The Industrial Medical Association becomes the American Occupational Medicine Association.

1977 Mine Safety and Health Administration is established to administer the provisions of the Mine Safety and Health Act of 1977.

American Association of Industrial Nurses is renamed as American Association of Occupational Health Nurses.

1980   First population-based and emergency room-based injury sur-
        veillance system is implemented in the United States (Massa-
        chusetts and Ohio).
1985   *Injury in America: A Continuing Public Health Problem* is published
        by the National Research Council and the Institute of Medicine.
1988   The American Academy of Occupational Medicine and the
        American Occupational Medical Association merge to become
        the American College of Occupational Medicine.
        Occupational Safety and Health Administration hires its first oc-
        cupational health nurse.
        *Role of the Primary Care Physician in Occupational and Environmental
        Medicine* published by the Institute of Medicine.
1991   *Disability in America: Toward a National Agenda for Prevention* is
        published by the Institute of Medicine.
        *Addressing the Physician Shortage in Occupational and Environmental
        Medicine* is published by the Institute of Medicine.
1992   Americans with Disabilities Act is passed.
1993   *Injury Control in the 1990s: A National Plan for Action* is published
        by the Centers for Disease Control.
1998   American Association of Occupational Health Nurses Founda-
        tion established.
1999   *Reducing the Burden of Injury* is published by the Institute of Medi-
        cine.

D
Locations of the National Institute for Occupational Safety and Health's Education and Research Centers (ERCs) and Training Program Grants (TPGs)

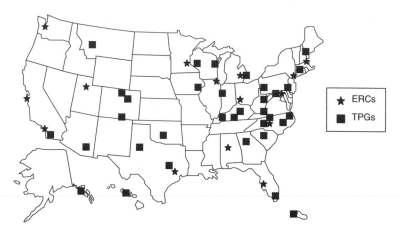

ERCs

Deep South Center for Occupational Health and Safety
 University of Alabama at Birmingham
 Auburn University
Northern California Center for Occupational and Environmental Health
 University of California at Berkeley
 University of California at Davis
 University of California, at San Francisco
Southern California Center for Occupational and Environmental Health
 University of Southern California
 University of California at Irvine
University of Cincinnati Education and Research Center
Harvard University School of Public Health Education and Research Center
Great Lakes Center for Occupational and Environmental Health
 University of Illinois, Chicago
 University of Illinois, Urbana
 Cook County Hospital
Johns Hopkins University Education and Research Center
University of Michigan School of Public Health Center for Occupational Safety and Health Engineering

Midwest Center for Occupational Health and Safety,
 University of Minnesota School of Public Health,
 Regions Hospital
University Occupational Safety and Health Education and Research Center
 Mt. Sinai School of Medicine, New York University
 Hunter College
 Environmental and Occupational Health Sciences Institute,
 UMDNJ-Robert Wood Johnson Medical School and Rutgers,
 The State University of New Jersey
University of North Carolina Occupational Safety and Health Education and Research Center
 University of North Carolina Chapel Hill School of Public Health
 Duke University
Sunshine Education and Research Center, University of South Florida
Southwest Center for Occupational and Environmental Health,
University of Texas Health Science Center at Houston
Rocky Mountain Center for Occupational and Environmental Health,
University of Utah
Northwest Center for Occupational Health and Safety, University of Washington

TPGs

Alaska Marine Safety Educational Association
University of Arizona
Catonsville Community College
Central Maine Technical College
University of Colorado
Colorado State University
University of Connecticut Health Center
Duke University Medical Center
East Carolina University
Emory University
University of Hawaii
University of Iowa
University of Kentucky
University of Massachusetts-Lowell
University of Massachusetts Medical Center
University of Miami, FL
Montana Tech of the University of Montana
Murray State University
North Carolina A&T State University
University of Oklahoma Health Sciences Center

University of Pennsylvania
University of Pittsburgh
University of Puerto Rico
Purdue University
St. Augustine's College
San Diego State University
University of South Carolina
Temple University
Texas A&M University
Texas Tech University
Trinidad State Junior College
Thomas Jefferson University
Virginia Polytechnic Institute and State University
Wayne State University
West Virginia University
Western Kentucky University
University of Wisconsin-Stevens Point
University of Wisconsin-Stout
Yale University